Beyond Safety Accountability

How to Increase Personal Responsibility

E. Scott Geller

Government Institutes
Rockville, Maryland

ABS Consulting

Government Institutes
4 Research Place, Rockville, Maryland 20850, USA
Phone: (301) 921-2300
Fax: (301) 921-0373
Email: giinfo@govinst.com
Internet: http://www.govinst.com

Copyright © 2001 by Government Institutes. All rights reserved.

04 03 02 01   5 4 3 2 1

No part of this work may be reproduced or transmitted in any form or by any means, electronic or mechanical, including photocopying, recording, or the use of any information storage and retrieval system, without permission in writing from the publisher. All requests for permission to reproduce material from this work should be directed to Government Institutes, 4 Research Place, Suite 200, Rockville, Maryland 20850, USA.

The author and publisher make no representation or warranty, express or implied, as to the completeness, correctness, or utility of the information in this publication. In addition, the author and publisher assume no liability of any kind whatsoever resulting from the use of or reliance upon the contents of this book.

ISBN 0-86587-893-5

Printed in the United States of America.

Table of Contents

Foreword 1

How To Use This Book 7

Chapter 1 — The Foundation — A Review of Behavior-Based Safety Principles 9

Chapter 2 — Safety Accountability Versus Responsibility 21

    A Personal Story 22

        From Failure to Synergy 22

        Relevance to Occupational Safety 23

    Actively Caring for Safety 25

        Three Ways to Actively Care 25

        Caring Directly or Indirectly 26

        An Illustrative Story 27

Chapter 3 — Decrease Top-Down Controls 31

    Perceptions of Individual Freedom 31

    Instructive Research 34

    Eliminate Fault Finding 34

        Attribution Errors 35

        Promote Fact Finding 35

    The Errors of Punishment 38

    The Information Processing Cycle 38

        Input Stage 39

  Interpretation & Decision-Making Stage 39

  Output Stage 39

 Slips vs. Mistakes 40

  Capture Errors 40

  Description Errors 40

  Loss-of-Activation Errors 41

  Mode Errors 41

 The Role of Experience 42

 Taking Responsibility for Our Errors 43

 Punishment and Accountability 43

  Unconscious Incompetence 45

  Conscious Incompetence 45

 The Futility of Punishment 48

 Context and Accountability 48

 Context at Work 50

Chapter 4 — Increase Feelings of Empowerment 53

 Hold People Accountable for Numbers They Can Control 55

 Outcome-Based Accountability 56

  A Subjective Indicator 57

  Inconsistent Reporting 57

  Losing Perceived Control 59

 Process-Based Accountability 60

  Safety Goal-Setting 61

Setting Goals Incorrectly 61

Set SMART Goals 61

Practical Examples 62

Vision vs. Goal 62

Principle of Shaping 67

Develop a Comprehensive Accountability System 68

What to Measure? 69

Evaluating Environmental Conditions 70

Evaluating Work Practices 70

Evaluating Person Factors 71

Accountability for Near Misses 74

Accountability for Property Damage 75

The Heinrich Ratio 75

Chapter 5 — Help People Feel Important 77

Increase Opportunities for Choice 78

A Laboratory Study 78

From Laboratory to Classroom 79

Practical Implications 80

Choice for Corporate Safety 81

From Choice to Importance 82

A Personal Story 84

What is Emotional Intelligence? 84

Intrapersonal IQ 86

Interpersonal IQ 87

Safety, Emotions, and Impulse Control 87

Nurturing Emotional Intelligence 88

Chapter 6 — Cultivate Belonging and Interpersonal Trust 91

Increasing Belonging 92

Actively Caring States 93

Understanding Interpersonal Trust 94

    Why the Resistance? 94

    What is Trust? 95

How to Build a Trusting Culture 97

    Communication 97

    Caring 98

    Candor 98

    Consistency 99

    Commitment 100

    Consensus 100

    Character 100

Chapter 7 — Teach and Support Safety Self-Management 103

What is Safety Self-Management? 105

From Outside to Inside Direction 106

From Outside to Inside Motivation 106

The Basics 107

Getting Started in Safety Self-Management 108

Safety as a Value	111
An Illustrative Anecdote	113
Inconsistent Behavior	113
What Consequences?	113
The Connection.	114
The Techniques of Safety Self-Management	114
Self-Observing and Recording	115
Developing a Self-Observation Checklist.	115
Using a Self-Observation Checklist	116
Intervening for Self-Improvement	117
Activator Management	118
Self-Statements	119
Self-Instruction	119
Beliefs	120
Interpretations	121
Mental Imagery	122
Self-Rewards	123
Goal-Setting	125
Social Support	126
Commitment	126
A Case Study	128
Implications	131

Bibliography 137

Glossary 141

Beyond Safety Accountability

Foreword

Sometime in the neighborhood of 1960-1961, I discovered the concept of accountability. I had been in safety for about six years doing what safety people did then. I inspected plants to find code violations (it was pre-OSHA). I harped on housekeeping. I presented and sat in on safety meetings, etc.

When reading an article in a management magazine I came across what I then perceived to be the key to it all — the key to safety excellence. That key was "accountability." If we can just pin it all down — define roles in terms of activities and tasks, ensure people at all levels are able (through knowledge and skill) to carry out those defined tasks, ensure that they in fact carry out those tasks through valid measurement, and ensure they continue to carry out those activities for an appropriate reward structure. If we can do this, then safety excellence will occur.

That was in the 1960's. I then went back to the drawing board in the early 1970's to search for other "keys" to safety performance. Besides the "management key," I discovered a "behavioral key" in organizational psychology, which said that through "motivation theory" and "positive reinforcement theory," we could enhance our engineering and our management approaches toward safety excellence.

I've pushed a management approach to safety for almost the last 40 years; and I've pushed a behavioral approach to safety for a little over 25 years. I believe the two go together. I usually am trying to convince managers and engineers to adopt a "behaviorally-sound" approach to safety. At a recent ASSE symposium I attempted to convince psychologists not to forget the management and engineering approaches to safety.

In all of this I've believed that accountability was the key. I've believed that stating that a person is responsible for safety is not enough. Instead, we must also hold that person accountable (through measurement and reward) in order to ensure proactive performance.

I've personally only known Scott Geller for a few short years. I've enjoyed his presentations at many conferences, and personally enjoyed him "off-stage" in one to one discussions. I've learned that for all of Dr. Geller's flamboyant stage presence, he is the most solid, research-based "guru" of today's "behavior-based" safety teachers.

Dr. Geller's theme in this book is somewhat different from mine. Whereas I say "responsibility" assignments are meaningless without accountability structures, Dr. Geller suggests that "accountability" is not enough — that those people who are held accountable must go further — they must feel a personal responsibility not only to fulfill their defined accountabilities, but also feel personally responsible to go "beyond" the measures of safety performance—to do more—to be personally responsible. He is right!

This book tells us how.

Dan Petersen, Ed.D, P.E., CSP
Fellow, American Society of
Safety Engineers

Acknowledgments

It seems I've been preparing all my life to write a practical set of books on how to make a difference in safety and health. So I'm indebted to a vast number of individuals — family members, teachers, researchers, colleagues, and students — who have guided me to this point. Many principles and procedures were developed and validated from our field research, which extended over 25 years and involved the collection and analysis of literally millions of behavioral observations by over a thousand university students. My graduate students managed most of this data collection, and I'm truly grateful for their valuable talents and loyal efforts.

Financial support from a number of corporations and government agencies made our 25 years of intervention research possible. We received significant research funding from the Alcohol, Drug Abuse, and Mental Health Administration; the Alcoholic Beverage Medical Research Foundation; Anheuser-Busch Companies, Inc.; Centers for Disease Control and Prevention; Domino's Pizza, Inc.; Exxon Chemical Company; General Motors Research Laboratories; Hoechst-Celanese; the Motor Vehicle Manufacturers Association; Motors Insurance Corporation; the National Highway Traffic Safety Administration; the National Institute on Alcohol Abuse and Alcoholism; the National Institute for Occupational Safety and Health; the National Science Foundation; Travelers Property Casualty; the U.S. Department of Education; the U.S. Department of Energy; the U.S. Department of Health, Education, and Welfare; the U.S. Department of Transportation; the Virginia Departments of Agriculture and Commerce, Litter Control, Motor Vehicles, and Welfare and Institutions. Profound knowledge is only possible through programmatic research, and these organizations made it possible for my students and me to develop and systematically evaluate ways to improve behaviors and attitudes throughout organizations and communities.

I am also indebted to the numerous guiding and motivating communications I've received from corporate and community safety professionals worldwide. Daily contacts with these people shaped my research and scholarship, and challenged me to improve the connection between research results and practical application. They also provided valuable positive reinforcement to prevent burnout. It would take pages to name all of these friends and acquaintances, and then I would necessarily miss many. You know who you are — Thank you!

The advice, feedback, and friendship of two people — Harry Glaser and Dave Johnson — have been invaluable for my preparation to write this book. My long-term alliance and synergism with these journalists has continuously improved my ability to communicate the practical implications of academic research and scholarship.

The drawings were created by George Wills (Blacksburg, Virginia). I think they add vitality and fun to the written presentation. I hope you agree. But without the craft and dedication of Mike Rowe, the illustrations could not have been electronically scanned for use by the publisher. In fact, Mike coordinated the final processing of each chapter, combining tables and diagrams (which he refined) with George Wills' illustrations, as well as the word processing from my dedicated and talented secretary — Gayle Kennedy.

I also sincerely appreciate the daily encouragement and inspiration from my current graduate students — Ted Boyce, Steve Clarke, Jason DePasquale, Julien Guillaumot, Chuck Pettinger, Mary Ann Timmerman, and Josh Williams; the administrators of our Center for Applied Behavior Systems — Kent Glindemann, Paul Michael, and Michael Rowe; and my colleagues at Safety Performance Solutions: Susan Bixler, Earl Blair, Anne French, Mike Gilmore, Betty Hillis, Molly McClintock, Sherry Perdue, and Steve Roberts. I'm also indebted to Steve Clarke's special advice and inspiration regarding preparation of the final chapter of this book on self-management techniques.

Plus, I'm grateful for the "actively caring" attention and support I've received from my friends and partners at J. J. Keller and Associates, especially Keith Keller and Susan Wrchota for pursuing the vision of our win/win collaboration, Geralyn Richards and Dick Ziebell for operationalizing the vision in the development and production of the videotape and manuals accompanying this book, Travis Rhoden for contributing to the vision by guiding the preparation, refinement, and production of this print material, and Scott Rogers and Daryl Brown for collaborating in the development of a seminar series to teach participants how to get the most from this product. Geralyn Richards deserves special recognition for overseeing this entire project and helping me handle the stress of deadlines.

All of these people, plus many, many more, have made possible this practical guide to developing the kind of accountability system for safety that can build personal responsibility. I am eternally grateful. I'm convinced the synergy from all your contributions will make it possible for readers to use this book and accompanying videotape to make their work environment a safer and more pleasurable place to spend their weekdays.

E. Scott Geller

The behavior-based safety methods described in this module have been successfully implemented by numerous Fortune 100 organizations in partnership with Dr. E. Scott Geller and his associates at Safety Performance Solutions, Inc.

Should you desire additional assistance implementing any of the processes described in these materials, Government Institutes is pleased to recommend Safety Performance Solutions, 1007 N. Main St., Blacksburg, VA 24060 (540) 951-7233.

How To Use This Book

Introduction

Beyond Safety Accountability is an educational tool designed to help you understand accountability from a behavior-based perspective. It presents methods you can use to implement and manage a behavior-based accountability system. This product is also an employee training tool. It provides practical tools employees can use to build personal responsibility.

What will this book do for me?

Learning Objectives

When you have completed this book on safety accountability you'll be able to:

- Distinguish between accountability and responsibility for safety, and explain why safety accountability is not enough.

- Explain why injury outcome statistics like TRIR (total recordable injury rate) are ineffective measures of safety performance.

- Classify actively caring behaviors as direct or indirect and as environment-focused, behavior-focused, or person-focused.

- Define empowerment from a person-state perspective and implement methods to increase feelings of empowerment.

- List problems in using top-down controls like punishment procedures to manage safety.

- Distinguish between outcome-based and process-based accountability, and explain why the latter is more important in building personal responsibility for safety.

- Define four types of errors or unintentional at-risk behaviors and explain how they are typically caused.

- Distinguish between errors, mistakes, and calculated risks and explain why punishment or "discipline" is rarely effective at decreasing these at-risk behaviors.

- Explain how the context in which we perform influences our attention to environmental hazards and our motivation to take a calculated risk.

- Distinguish between vision and goal, and explain the role of each in developing a safety accountability system that promotes individual responsibility.

- Set effective safety goals that motivate involvement and build commitment and responsibility.

- Use the principle of shaping to increase safety-improvement behaviors.

- Explain why a causal analysis of property-damage incidents needs to be included in a safety accountability system.

- Clarify why it's important to include opportunities for individual and group choice in a safety accountability system.

- Increase employees' feelings of importance with regard to their safety-improvement efforts.

- efine emotional intelligence and explain its relevance to industrial safety and health promotion.

- Develop a comprehensive safety accountability system that increases employee involvement and builds personal responsibility for safety.

- Implement a behavior-based self-management process that increases your personal accountability and responsibility for safety.

- Teach safety self-management to others so they can improve their personal control of their own safety-related behaviors.

- Establish various support mechanisms for safety self-management.

The Foundation: A Review of Behavior-Based Safety Principles

Chapter 1

Taking Control of Safety

When people take control of safety, they see safety and health performance improve and they feel good about it, because they know they're making a real difference. They get a sense of accomplishment that keeps them going. When I describe the mission of involving employees in achieving a Total Safety Culture, you know what I often hear from company managers? "Right, Doc. That sounds good, but I can't imagine our people caring that much about safety and health." Quite a few of you may harbor the same skepticism.

Given the way most companies have been dealing with safety and health, total employee involvement for occupational safety can be hard to imagine, let alone achieve. But people really do care about their own health and safety, and most of us also care about other people's safety. No one wants to see another person get hurt. People just don't know what to do to prevent injuries. They don't know how to act on the fact that they care. So we need to teach them how to act on their caring—how to actively care.

That's what this book is all about. It shows you how to implement a behavior-based accountability process that allows you to take control of safety and keep people from being injured. When you complete this book, you will have effective methods and tools to use to make a big difference in occupational safety and health.

Stop Thinking Safety Program

Safety programs traditionally have been top down. What often happens is that the safety person goes out and finds a program. People get trained, they watch videos, and they're told exactly what to do. And then the program doesn't work, or it doesn't work well enough. Maybe some improvement occurs for awhile, but then things start to slide back to where they were. So what do we do? We try another program, and the same thing happens. And pretty soon you lose employee buy-in because employees begin to believe the safety program is just a new flavor of the month.

Now, I'm not saying we haven't made progress in safety and health. Of course we have. The workplace today is a lot more safe than it was 50 years ago. What I am saying is that we're not making good progress TODAY. Even in a model facility, a place where safety performance is exemplary, safety has plateaued. Safety performance is not improving. Too many people are still getting hurt.

Safety Is a Continuous Process

If we hope to improve safety performance we need to stop thinking "safety program," and start thinking "safety process." A program sounds like it's set in concrete. It sounds like safety is just one part of your job. You've got your quality program and you've got your safety program. But safety isn't like that, or it shouldn't be anyway.

Safety is really a process, and that means it's continuous. If we really want to reduce work-related injuries, and keep reducing them, we need to make safety a way of life. We need to involve people in daily activities consistent with the vision of a Total Safety Culture. We start by teaching people the principles behind safety improvement. Then we let THEM apply those principles and in turn control the process. People on the shop floor have the most opportunity to control safety. They're the ones who can have the biggest impact on keeping people safe — if they understand the principles behind the safety process and if they get involved. This book shows you how to make that happen with an accountability system that builds personal responsibility for safety. We're talking about safety by the people and for the people. That's what's going to work.

Common Sense vs. Science

The processes described in this book address the human dynamics of occupational safety. Reducing injuries below current levels requires increased attention to human factors. The engineers and policy makers have made their mark. Now it's time to deal with the human dynamics of injury prevention — the psychology of safety.

Many people, including safety professionals, believe psychology is just basic common sense. For these individuals, it seems appropriate to do just whatever comes naturally when dealing with the human dynamics of safety. We have learned a lot about the psychology of human behavior and attitudes throughout our lives. But this learning has not been representative nor objective.

A psychology of safety must be based on rigorous research, not common sense or intuition. That's what science is all about. Much of the psychology in self-help paperbacks, audiotapes, and motivational speeches is not founded on scientific investigation, but is presented because it sounds good and will "sell." The principles and procedures in this book were not selected on the basis of armchair hunches but rather from the relevant research literature.

We do not want engineers to construct buildings with only common sense, nor do we want surgeons to operate on family members with only good common sense. Rather, we expect these professionals to have appropriate knowledge and skills, founded on reliable and valid findings from objective research. Likewise, science provides profound knowledge and direction for dealing with the human dynamics of safety. Without such direction, we rely on only what sounds good to our biased ears.

A "Hot" Topic

Behavior-based safety is a "hot" topic these days at company safety meetings and professional development conferences. Safety professionals readily relate to the term "behavior," because they realize that human behavior is involved to some extent in almost every environmental hazard and personal injury. So improving behavior is key to preventing injuries. In other words, safety leaders and consultants recognize that unsafe or at-risk behavior is a frequent cause of both minor and major injuries, and therefore substituting safe for at-risk behavior is critical for an upstream proactive approach to industrial health and safety. Unfortunately, there is much distortion and confusion among safety professionals and consultants regarding behavior-based safety. There is much more to behavior-based safety than believing that behavior change is critical for injury prevention.

The Narrow Perspective

Many have a narrow perspective regarding behavior-based safety, considering it only a tool or program to manage behavior. It often gets portrayed as the narrow set of procedures marketed by a certain consulting firm to improve safety-related behaviors. For example, many seem to believe behavior-based safety is nothing more than the use of checklists of critical behaviors. To these individuals, behavior-based safety is merely employee observation and feedback.

Self-Destructive Myths

In this book I hope to dispel the self-destructive myths associated with narrow and misinformed impressions of behavior-based safety. I'll explain how the principles of behavior-based safety can be used to develop an objective accountability system. And, I'll show how to implement an accountability system so people develop personal responsibility for safety. Combining the principles of behavior-based safety with concepts of accountability and personal responsibility puts you on the road to achieving a Total Safety Culture.

The Seven Principles of Behavior-Based Safety

Other books I have written (Geller, 1996a, 1997c) detail the principles and procedures of the behavior-based approach, and show why this approach to motivating safety improvement reflects actively caring. They illustrate how to implement behavior-based safety in ways that build personal commitment, group cohesion, and various other internal dimensions of people that increase their willingness to actively care for the safety and health of others.

Obviously, I cannot review the contents of those earlier books here, but I can review seven basic principles of behavior-based safety. These seven principles should serve as criteria to follow when developing a behavior-based tool or method for safety management. The seven principles are broad enough to encompass a wide range of practical operations, but are narrow enough to guide the development of effective methods for managing the human dynamics of health and safety. I propose these as a mission statement or map against which to check your efforts to improve safety-related behaviors and attitudes. It will become obvious that these principles are applicable to continuous improvement challenges beyond occupational safety and health.

1. Focus on Observable Behavior

B.F. Skinner laid the groundwork.

The behavior-based approach to safety is founded on behavioral science as conceptualized and researched by B. F. Skinner (1938, 1953, 1974). Experimental behavior analysis, and later, applied behavior analysis, emerged from Skinner's research and teaching. He laid the groundwork for numerous therapies and interventions to improve the quality of life of individuals, groups, and entire communities (Goldstein & Krasner, 1987; Geller, Winett, & Everett, 1982; Greene et al., 1987). Whether working one-on-one in a clinical setting or with work teams throughout an organization, the intervention procedures always target specific behaviors in order to produce constructive change. In other words, the behavior-based approach focuses on observing what people do, analyzes why they do it, and then applies a research-supported intervention strategy to improve what they do.

Whatever the intervention strategy used to improve a human aspect of safety, the process should target behavior. Whether using training, feedback, incident analysis, coaching, or incentives to benefit safety, focus on behavior. Why? Well, first you can be objective and impersonal about behavior. You can talk about behavior independently from people's opinions, attitudes, and feelings. Behavior varies according to factors in the external world, including equipment design, the management system, the behaviors shown by others, and various social dynamics. An open discussion about the environmental and interpersonal determinants of safe versus at-risk behavior can lead to practical modifications of the work culture to encourage safe behavior and discourage at-risk behavior.

Act people into different thinking!

Behavior-based intervention "acts people into thinking differently," whereas person-based intervention "thinks people into acting differently." The person-based approach is used successfully by many psychiatrists and clinical psychologists in professional therapy sessions, but it is not cost-effective in a group or organizational setting. To be effective, person-focused intervention requires extensive one-on-one interaction between a client and a specially trained intervention specialist. Even if time and facilities were available for an intervention to focus on internal and nonobservable attitudes and person states, few safety professionals or consultants have the education, training, and experience to implement such an approach. Internal person factors can be improved indirectly, however, by directly focusing on behaviors in certain ways.

The key is to focus on behavior and you'll be on the right track. Actually, whatever the intervention approach, focus on behavior. It's behavior-based commitment, behavior-based goal-setting, behavior-based feedback, behavior-based training, behavior-based incident analysis, behavior-based recognition (Geller, 1997b), behavior-based incentives and rewards (Geller, 1996b), and behavior-based teamwork (Geller, 1998).

2. Look for External Factors to Understand and Improve Performance

Focus on behavior and external causes.

Internal person dimensions like attitudes, perceptions, and cognitions are difficult to define objectively and change directly. So stop trying! Most of us don't have the education, training, experience, nor time to deal with people's attitudes or person states directly. Instead, target behaviors caused by external system factors independent of individual feelings, preferences, and perceptions. When you empower people to analyze behavior from a systems perspective and to implement interventions to improve behavior, you will indirectly improve their attitude, commitment, and internal motivation.

In the first widely-used American textbook in psychology, *Principles of Psychology*, William James (1890) explained the reciprocity between behavior and attitude as follows:

> Sit all day in a moping posture, sigh, and reply to everything with a dismal voice, and your melancholy lingers . . . If we wish to conquer undesirable emotional tendencies in ourselves, we must . . . go through the outward movements of those contrary dispositions which we prefer to cultivate.

Internal causes are difficult to identify and change directly.

Careful observation and analysis of ongoing work practices can pinpoint many potential causes of safety and at-risk behaviors. Those causes external to people — including management systems, policies, or supervisory behaviors — can often be altered for the improvement of both behavior and attitude. In contrast, internal person factors are difficult to identify, and if defined, they are even more difficult to change directly. So with behavior-based safety the focus is placed on external factors — environmental conditions and behaviors — which can be changed before a potential injury.

3. Direct with Activators and Motivate with Consequences

This principle explains why behavior occurs, and is used as a guideline for developing intervention strategies to change behavior in desired directions. It actually runs counter to common sense. When people ask us why we did something, we give reasons like, "Because I wanted to do it," "Because I needed to do it," or "Because I was told to do it." These reactions seem to put the cause of our behaviors before their occurrence. And this notion is generally supported by the variety of "pop psychology" self-help books and audiotapes, which claim we motivate ourselves with self-affirmations, positive thinking, optimistic expectations, or challenging intentions.

Behavior is directed by activators.

The fact is, however, we do what we do because of the consequences we expect to get for doing it. As Dale Carnegie put it in his classic best seller *How to Win Friends and Influence People*, "Every act you have ever performed since the day you were born was performed because you wanted something" (p.62). It's noteworthy that Carnegie (1936) referred to the research and scholarship of B. F. Skinner as the foundation of this motivation principle.

The important point here is that activators (or the signals preceding behavior) are only as powerful as the consequences supporting them. In other words, activators tell us what to do in order to receive a consequence, from the ringing of a telephone or doorbell to the instructions from a training seminar or one-on-one coaching session. However, we will follow through with the particular behavior activated (from answering a telephone or door to following a trainer's instructions) to the extent we expect doing so will give us a pleasant consequence or help us avoid an unpleasant consequence.

Behavior is motivated by consequences.

This principle is typically referred to as the ABC model or three-term contingency, with "A" for activator, "B" for behavior, and "C" for consequence. More than 40 years of behavioral science research has demonstrated the effectiveness of using this ABC model to design interventions for improving behavior at individual, group, and organizational levels. The next principle provides more specific direction for improving safety-related behaviors.

4. Focus on Positive Consequences to Improve Behavior

The most powerful consequences are soon, certain, and significant.

I've heard behavior-based consultants use the phrase "soon, certain, and positive" to describe the most powerful motivating consequences. These same consultants relate "soon" to "timing," "certain" to "probability," and "positive" to "significance." This instruction is accurate with regard to the first two characteristics of consequences. The most powerful motivating consequences are "soon" and "certain." That's why much at-risk behavior occurs. Compared to safe behavior, at-risk behavior provides "soon" and "certain" consequences such as comfort, convenience, ease, and faster job completion.

The consequences mentioned above as motivators for at-risk behavior are all positive. But, it is inappropriate to consider "positive" in the realm of "significance." A negative consequence like property damage or personal injury is certainly significant. It's just that the significance of such a negative consequence is lost because of its remoteness and low probability of occurrence. Note, however, the certainty and probability of a significant negative consequence is only low when we take a narrow individualistic perspective. From a large-scale systems perspective, property damage and injuries are not so remote and infrequent. Property damage and injuries are happening out there to people in our culture, and we care about that — right? Thus, we need to teach and encourage systems thinking and interdependency throughout a work culture.

Positive consequences improve behavior and attitude.

As this third principle indicates, using positive over negative consequences is critically important, but it's not relevant to "significance." It's relevant to "attitude," and many other internal dimensions of people. Think about it. How does a reward, personal recognition, or a group celebration make you feel, compared to a reprimand or criticism? Both consequences are significant with regard to behavioral impact. The difference is in the accompanying attitude or feeling state.

We learn more from our successes than our failures.

As I've described in the October 1997 issue of *Professional Safety* (Geller, 1997b), when positive recognition is delivered correctly, it not only increases the frequency of the behavior it follows, it also improves morale, attitude, and various other person states. This in turn increases the likelihood other safe behaviors will occur, and that positive recognition will be used more often to benefit both behavior and attitude.

Contrary to the views of some "pop psychologists," we learn more from our successes than our failures. So recognizing people's safe behavior will facilitate more learning and positive motivation than will criticizing people's at-risk behavior. Remember that only with positive consequences can you improve both behavior and attitude at the same time. But without an objective and systematic evaluation process, we can't be sure our interventions have the beneficial effects we want. This brings us to the next basic principle of behavior-based safety.

5. Apply the Scientific Method to Improve Intervention

Learn from the scientific method.

Continuous improvement requires continuous evaluation. I'm sure you've heard the saying, "Feedback is the Breakfast of Champions." Well, that's the essence of this principle. The scientific method, not common sense, can provide the kind of objective feedback necessary for improving an intervention process.

Common sense is based on people's selective listening and interpretation, and as a result, it is necessarily biased. Common sense sounds good to the individual listener, but may not be accurate from an objective or systems perspective. The most effective intervention approach might not sound good to the individual listener, at least not at first.

Systematic and scientific observation enables the kind of objective feedback needed to know what works and what doesn't work to improve behavior. B. F. Skinner rejected unobservable inferred constructs (like intentions and attitudes) for scientific study because it is very difficult (if not impossible) to apply the scientific method to these internal constructs and obtain practical feedback for continuous improvement.

Learn continuously from the DO IT process.

Behavior can be objectively observed and measured before and after the implementation of an intervention process. This application of the scientific method provides the kind of feedback that can be used for continuous safety improvement. My associates and I use the acronym "DO IT" to teach this principle of behavior-based safety to employees who are empowered to intervene on behalf of their coworkers' safety and want to continuously improve their intervention skills.

D* = *Define
O* = *Observe
I* = *Intervene
T* = *Test

The four steps of the DO IT process are reflected by each letter: D = Define the target behavior to increase or decrease; O = Observe the target behavior during a preintervention baseline period to understand natural environmental or interpersonal factors influencing the target behavior (see Principles 1 & 2); I = Intervene to change the target behavior in desired directions; and T = Test the impact of the intervention procedure by continuing to observe and record the target behavior while the intervention is in effect.

With the DO IT process, an intervention can be objectively evaluated for unbiased decision making. Comparisons between observations taken during baseline and during the intervention might inform us to continue the intervention process, implement another intervention approach, or define another behavior for another run through the DO IT process. The systematic evaluation of a number of DO IT processes can lead to a body of knowledge worthy of integration into a theory. This possibility is reflected in the next principle.

Don't be blinded by theory.

6. Use Theory to Integrate Information, Not to Limit Possibilities

While much, if not most, research is theory driven, Skinner (1950) was critical of focusing the design of an experiment on the testing of theory. Theory-driven research can narrow the view of the investigator and thus limit the discovery of important findings. In other words, applying the scientific method merely to test a theory can be like putting blinders on a horse. It can limit the breadth of information gained from systematic observation.

Be open to many possibilities.

Many valuable findings about human dynamics have resulted from exploratory investigation. Many behavioral scientists have recorded systematic observations of behavior before and after an intervention or treatment procedure to answer the question "I wonder what will happen if we do this?" rather than "Is my theory correct?" These researchers were not looking for specific results, but were open to finding anything. Modification of their research designs or observation procedures were guided by their behavioral observations, not by a particular theory. Thus, their innovative research was data driven rather than theory driven.

This reflects an important perspective for safety professionals, especially when applying the DO IT process. It's often better to be open to many possibilities for improving safety performance than to focus on supporting a certain process. Numerous intervention methods are consistent with a behavior-based approach to safety, and an intervention process that works well in one situation will not necessarily be effective in another setting. So make an educated guess at what intervention approach to use at the start of a behavior-based safety process, but be open to results from a DO IT process and refine your procedures accordingly.

Use theory to integrate information.

After many systematic applications of the DO IT process, you will notice distinct consistencies. Certain procedures will work better in some situations than others, with some individuals than others, or with some behaviors than others. You might summarize relationships between intervention impact and specific situational or interpersonal characteristics. In this way you are developing a research-based theory of what works best under given circumstances. You are using theory to integrate information gained from systematic behavioral observation. Skinner (1950) approved of this use of theory, but cautioned that premature theory development can lead to premature theory testing and limited profound knowledge.

7. Design Interventions with Consideration of Internal Feelings and Attitudes

When working to avoid failure we feel controlled.

B. F. Skinner's concern for people's feelings and attitudes is reflected in his contempt for the use of negative consequences to motivate behavior. In his classic book, *Beyond Freedom and Dignity,* Skinner (1971) writes, "The problem is to free men, not from control, but from certain kinds of control" (p.41). Then he explains why control by negative consequences needs to decrease in order for people to feel free. Think about it. When do you feel personal freedom or empowerment — when you are working to achieve a pleasant consequence or working to avoid an unpleasant consequence?

When working to achieve we feel free.

Some of my university students are motivated to avoid failure (reflected in a poor grade), while other students are motivated to achieve success (as in a good grade or a worthwhile learning experience). Which students feel more in control of their class grade and thus have a better attitude toward my class? You know the answer to this question because you can reflect on your own feelings or attitude in similar situations.

The important lesson here is that we perform behavior to receive positive consequences or to avoid or escape negative consequences. And, the type of consequence (positive or negative) motivating our behavior affects our attitude. We can often intervene with others to increase their feelings of control, accomplishment, and personal freedom. Even the things we say to others, perhaps as a statement of genuine approval or appreciation for a job well done, can influence feelings of personal freedom and empowerment.

The rationale for using more positive than negative consequences to motivate behavior is based on the differential feeling states provoked by positive reinforcement versus punishment procedures. Similarly, the way we implement an intervention process can increase or decrease feelings of empowerment, build or destroy trust, or cultivate or stifle a sense of teamwork or belonging. Thus, it's important to assess feeling states or perceptions that occur concomitantly with an intervention process. This can be accomplished informally through one-on-one interviews and group discussions, or formally with a perception survey.

Therefore, decisions regarding which intervention to implement and how to refine existing intervention procedures should be based on both objective observations of behaviors and subjective evaluations of feeling states. Often, however, it's possible to evaluate the indirect internal impact of an intervention by imagining yourself going through a particular set of intervention procedures and asking the question "How would I feel?" Perhaps in this case, basic common sense is as good as any evidence you could gather from subjective evaluations of other persons' feeling states.

Do these principles sound like good old common sense to you? If so, I'm also pleased. But I must warn you that others will not necessary feel the same without appropriate education, training, and guided experience. Common sense or intuition is often not correct. What sounds good to one person will not necessarily sound right to another. Consider, for example, the following "common sense" strategies people have implemented quite often in an attempt to deal with the human dynamics of safety.

- Punish a person who returns to work after a lost-time injury.

Many common sense approaches to safety management are ineffective, and can be harmful.

- Establish a system whereby employees must observe one unsafe condition or behavior each day and "fix" it.

- Implement a safety incentive program whereby everyone in a work area gets a prize if no one reports an injury.

- Set up a "safety employee of the month" program in which one individual in a large facility is publicly recognized for having the "Best Safety Attitude."

- Invite a motivational speaker to address all employees with topics like "Try Harder," "Change Your Attitude About Safety," "Self-Affirmation is the Key to Motivation," or "Safety Awareness and a Positive Attitude are Key to Behavior Change."

- Post signs with slogans like "Think Safety," "Safety is a Condition of Employment," "Zero Accidents is Our Goal," "Safety is a Priority," or "All Injuries are Preventable."

Behavior-based safety is based on results from research.

All of these strategies are ineffective and run counter to the behavior-based principles described here. And some of these techniques can do more harm than good to the human dynamics of industrial safety and health. Yet, I'm sure you've seen, even experienced, some of these intervention approaches. Why? Because they seemed like good common sense to someone. It takes empirical investigation, not common sense, to guide the development and implementation of an improvement intervention, whether repairing a bridge, constructing a building, or administrating an incentive/reward process.

Behavior-based safety, as reflected in the seven principles described here, is based on 40 years of rigorous research. And with additional research, the methods and tools derived from behavior-based safety will continuously improve. In other words, with ongoing application of the scientific method we build and refine our common sense, and thus become more effective managers of safety-related behaviors and attitudes.

Achieving a Total Safety Culture: The Safety Triad

Achieving a **Total Safety Culture** is much easier said than done, but it is achievable if the three domains depicted in the figure below are addressed appropriately. Every aspect of safety falls under the three sides of the Safety Triad— Environment, Behavior, and Person. And to cultivate a Total Safety Culture, we need to pay attention to factors within each of these domains.

Environment refers to everything around us. It's there even when there are no people around. It includes the machines, buildings, tools, personal protective equipment, materials, and the temperature. It also includes things like safety policies and rules. These are as much a part of a company's safety environment as anything else.

Chapter 1 — A Review Of The Principals

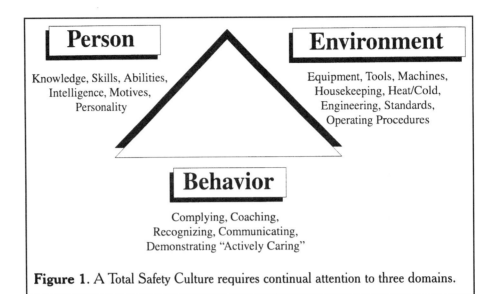

Figure 1. A Total Safety Culture requires continual attention to three domains.

Over the years, we've paid the most attention to environmental factors, and that's made a difference. We need to keep actively caring about the environment and improving workplace conditions. But we have not paid enough attention to people — the other two sides of the triangle.

Behavior refers to the things people actually do. It's the outside stuff, the stuff we can see and identify. You turn left or you turn right. You run or you walk. You put on safety glasses or you don't. In other words, behavior is observable and objective. For example, if I could see you right now, I might find you sitting at a desk reading this book. That's observable behavior. Other people watching you would see the same thing. On the other hand, I might see you lying on the floor with this book in front you. That's behavior too, because I could describe it objectively and accurately.

Behavior is objectively observable.

However, could I tell by observing you whether you're reading these words and comprehending their meaning? You might be sitting up in a chair, looking at this page and appear to be understanding my words. But you might be concentrating on what you're having for dinner tonight. Right? And what about your attitude? Could I tell how you feel about what I've written? An observer could not tell whether you understand, agree, or disagree with the message here.

Of course, a person could ask you whether you understand and agree with my words. You might think you understand something, but could not explain it very well to someone else. Or, you might disagree with a point, but not reveal your disapproval. We're talking about what's inside a person, and that can get pretty complicated.

Person factors are subjective.

Person factors are the inside stuff, including our knowledge, skills, abilities, attitudes, and feelings and values. Psychology deals with both the inside person factors and the outside behavior factors. Both need our careful attention throughout the cultivation of a Total Safety Culture.

Nearly every injury that happens involves at-risk behavior — the unsafe acts that people perform. So if we want to reduce injuries, we need to understand why people often act unsafely. What's the psychology behind ignoring safety and taking risks? Why do we do it? What can we do to reduce the number of at-risk behaviors that occur and increase the number of safe behaviors? You see, that's really the bottom line. Injuries will be reduced only if at-risk behaviors decrease and safe behaviors increase.

Four Phases of Implementation

The behavior-based approach to helping organizations achieve a Total Safety Culture typically follows a four-phase framework. Of course, the structure of training and implementation efforts in a given organization depends upon its size, culture, and specific safety improvement goals. Therefore, tailor your approach to best suit your organization's culture and needs.

During ***Phase 1***, you should assess the current safety culture in your organization and learn about the organization's infrastructure in order to plan an appropriate strategy for implementation. This phase may also include providing key members of the organization an overview of the principles and tools of behavior-based safety, as well as a general description of the training and implementation methods.

Phase 2 consists of training a team of employee representatives to serve as leaders and champions of the behavior-based safety process. Details about key behavior-based safety teams are covered in my recent book on teamwork (Geller, 1998).

During ***Phase 3***, all employees within the organization (or division) receive training regarding the principles, tools, and implementation methods. The team leaders and champions can lead this training.

Behavior-based safety needs champions and teamwork.

Finally, during ***Phase 4***, teams of employees implement and monitor behavior-based observation, feedback, and measurement efforts. This requires interdependent teamwork, along with various individual and group dynamics, including: interpersonal trust, team leadership, productive team meetings, team evaluation and accountability processes, team-based goal-setting and corrective-action planning, and a cooperative, win/win perspective.

Think people into different acting.

This book shows you how to develop and deliver an accountability system that can direct and motivate individuals to go beyond the call of duty to achieve an injury-free workplace — a Total Safety Culture.

Safety Accountability Versus Responsibility

Chapter 2

Safety accountability is not enough.

The other day I received a very insightful and provocative letter from Niru Davé, Manager, Safety and Training of Champion International. He said some very complementary things about my book "Working Safe," and included a study guide he had developed to facilitate its instructional applications. Then came the most thoughtful part of this communication. He asked me a few questions which reflect the most important and challenging behavioral problem in the safety and health field. Simply put, the issue is this: Once an effective behavior-based safety process is up and going, what can be done to facilitate a transition from compliance to the formal structure of a process with a built-in accountability system to informal, self-directed, and continuous contributions to safety which are not objectively counted or tracked? In other words, how can we get people to take their personal responsibility for safety beyond that which they can be held accountable for? Research suggests there are several things we can do. These things are presented in this book. Niru Davé set the stage for his questions about transitioning from accountability to responsibility by defining the following ideal **Total Safety Culture**:

> The heart of any behavior-based safety process and the concept of the Total Safety Culture is the fact that employees actively care for each other, keep eyes open to notice unsafe behaviors and provide immediate feedback to the person engaging in such behaviors. Also, occasionally, they notice safe behaviors and provide compliments (recognition). This happens all the time and continuously.

Behavior-based coaching provides an objective accountability system.

Then Niru presented the critical issue. He acknowledged that behavior-based safety is essentially a "crutch" that provides "structural arrangements and specific expectations for employees to engage in observation and provide (behavior-based) feedback." In other words, a behavior-based coaching process provides an accountability system for tracking individual participation, as well as the percentage of certain safe and at-risk behaviors. But while this objective and systematic process is an improvement over traditional top-down approaches to occupational safety, it is not ideal.

What can be done, if anything, to move an organization to the ultimate Total Safety Culture in which everyone actively cares for each other's safety? More specifically, what procedures, processes, or culture-change initiatives can facilitate the occurrence of actively-caring activities in between scheduled observation and feedback sessions? Employees can't be expected to fill out an audit card every time they see an environmental hazard or at-risk behavior, yet such situations call for corrective action.

Chapter 2 — Safety Accountability versus Responsibility

It takes more than accountability to achieve a Total Safety Culture.

Can we expect employees to actively care for safety when their actions are not formally counted or tracked in an accountability system? And, how should management be held accountable for supporting rather than stifling workers' efforts to do the right thing for their safety and health, which includes reporting and analyzing all incidents related to their safety and health? From my perspective, all of these questions boil down to understanding the difference between accountability and responsibility, and then implementing interventions that enable or motivate people to extend their personal responsibility for safety beyond that for which they are held accountable.

A Personal Story

At the age of ten I began taking drum lessons. My parents bought me a used snare drum and arranged for private lessons. Once a week my drum teacher came to my house and gave me 30 minutes of one-on-one instruction. Even though it happened more than 40 years ago, I can still remember how excited and motivated I was about this experience. My parents had promised to buy me a complete drum set (secondhand of course) if I showed continuous progress for one year. I was very motivated to make that happen.

In the beginning I looked forward to every drum lesson, because I could show progress. My teacher held me accountable for completing certain homework assignments, and with a minimal amount of practice, I could show him my accomplishments. I consistently answered his challenge of learning particular drum-beat patterns or correctly translating the rhythm notes of marching music he assigned from my "drum book." Often I achieved more than the assignment, thus extending personal responsibility beyond what my teacher held me accountable for. Whenever I did this I experienced special pride, self-confidence, and personal control.

From failure to synergy. During the fourth month of my first year of drum lessons I had an experience that almost resulted in me "hanging up my sticks." My teacher held me accountable for learning the drum roll. Besides demonstrating this basic percussion element, he didn't give me much advice about how to learn this relatively complex drum-stick maneuver. Each week he asked me to show him the drum roll and I couldn't do it. He was asking me whether I was accountable, and for many weeks I had to say "no." I just didn't get it. I couldn't figure out how to acquire this essential drumming skill.

Losing control saps empowerment.

My inability to master the drum roll decreased my sense of personal control in this situation. My empowerment was sapped. I no longer looked forward to my weekly drum lesson. I dreaded the teacher's accountability question, "Let's see your progress on the drum roll." I developed aversive feelings of low personal control or helplessness about this particular assignment. If there was a way to cheat or beat the system, I would have done it. My learning on the drum had plateaued, and I began saying to myself, "I can't do this."

The incentive of having a complete drum set kept me going. A discussion with the music teacher at my elementary school was critical in enabling me to learn the drum roll. He explained that the drum roll is really nothing more than two beats with each stick, accomplished very quickly and fluently. He broke down my overwhelming assignment into simple, manageable parts I could handle. I began approaching the drum roll with this new paradigm, eventually developing fluent and smooth two-beat control per hand.

Regain control by breaking down complex tasks into manageable parts.

After several hours of practicing the two-beats-per-stick routine, it happened. I was suddenly doing a drum roll. It seemed like a miracle at the time. The two-beats-per-stick suddenly became indistinguishable and a seemingly complex drum roll emerged. The whole seemed much greater than the sum of its parts. Today I reflect on that as an elegant demonstration of synergy.

That was certainly not the only threat to my eventual participation in all-state band and orchestra, and ten-year career as a professional rock-and-roll drummer. But subsequent challenges benefited from my remembering I got through that crisis by breaking down the whole (the outcome goal) into manageable parts (or process goals). Often the accountability system did not acknowledge the process goals or give credit for accomplishing them. Thus, it was necessary to adopt some self-management techniques, which essentially consisted of giving myself internal direction and rewards. I consider this process "self-discipline." But in those good-old days, "discipline" did not have the negative and punitive connotations it has today.

Relevance to occupational safety. I hope you see the relevance of this childhood memory to occupational safety, particularly the challenge of motivating people to do more than what's required. I'm sure you've experienced similar events in your life. How many times have you been asked to do something which seemed insurmountable at first? Perhaps you lacked the knowledge, skills, or resources deemed necessary to meet the challenge. Perhaps the outcome you were held accountable for required a synergistic combination of several activities, but you weren't sure what these activities were. Or, perhaps you knew what behaviors were necessary to achieve the outcome goal, but you didn't know how to get all of these behaviors to occur. If you've been overwhelmed by a teacher's or supervisor's assignment, you've also experienced feelings of helplessness or loss of personal control. In such cases, you've also encountered a deficit in personal responsibility for a particular goal, purpose, or mission statement. When this happens, personal responsibility and accountability are not the same.

Is your accountability for safety and responsibility for safety the same?

Do you perceive your personal responsibility for safety to be about the same as your accountability for safety? I've met many safety leaders who would answer "yes" to this question. They explain how their manager or CEO holds them accountable for maintaining or achieving a low injury rate, and how they feel constant pressure or stress to do whatever it takes to keep the incident numbers down. Often a significant financial bonus is contingent on maintaining a low rate of OSHA recordables. Sometimes, the financial bonus of several individuals depends upon a single number — the total recordable injury rate (TRIR) at a certain industrial site.

Holding people accountable for the wrong numbers can reduce personal responsibility.

When it comes to safety there seems to be a readily available accountability system in place. The company safety department tracks frequency and severity of employee injuries, as well as the amount of money paid for workers' compensation. Since every injury detracts from corporate profits, it seems reasonable to link a person's financial bonus to injury rate. Better yet, factor in the severity of the injury and you have a seemingly very fair system. The less money a company pays for industrial injuries, the more financial profit and the more money available for an end-of-the-year bonus. This seems like good common sense, and is in fact an accountability system used by many companies to motivate top managers to put more "effort" into safety.

Is there a downside or disadvantage to holding people accountable for the frequency and/or severity of employee injuries? You bet there is. Depending upon who is held accountable, the nature of the rewards or penalties, and most importantly, the kind of safety improvement system put in place to influence the injury records. Research on bonus pay systems has shown motivational advantages to linking financial rewards to performance variables. However, serious problems arise when the people held accountable for certain numbers don't know how to influence those numbers. It's analogous to the youngster being held accountable for a drum roll, but not knowing how to perform the component parts necessary for a synergistic outcome.

Safety accountability is not enough. Even an appropriate accountability system is not sufficient to achieve the Total Safety Culture required for an injury-free workplace. As depicted in Figure 2, people's personal responsibility for safety needs to extend beyond accountability. This cannot happen when people are held accountable for numbers they don't believe they can control. As long as I believed mastery of the drum roll was outside my domain of influence, I made no progress. But when I learned the right process goals needed for the outcome goal, my involvement escalated and led to increased commitment and responsibility.

Vision can motivate responsibility.

The vision of an ultimate reward (a complete drum set) was invaluable in maintaining my self-motivation to achieve the many small wins needed for continuous improvement. Likewise, the vision of an injury-free workplace keeps safety leaders motivated to establish the many conditions needed to help people extend their personal responsibility for

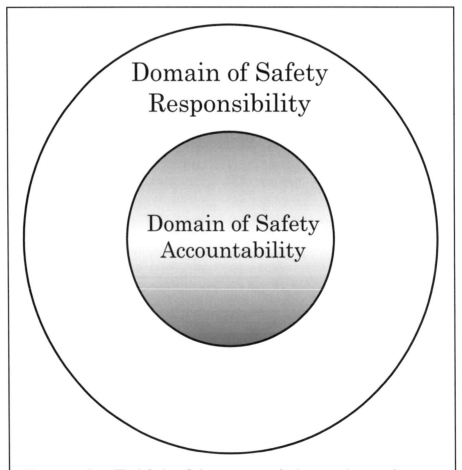

Figure 2. In a Total Safety Culture, everyone's domain of personal responsibility for safety is greater than their domain of safety accountability.

safety beyond the accountability system. As I've written in other sources (Geller, 1996a, 1997c), an injury-free environment or a Total Safety Culture requires people to go beyond the call of duty for the safety and health of themselves and others. "Beyond the call of duty" means doing more than expected or what an accountability system calls for. In a Total Safety Culture, everyone feels responsible for safety and continually looks for ways to act on their responsibility.

Outcome-based accountability provides no direction, and can stifle personal control.

An accountability system that focuses only on ultimate outcomes does not provide direction or incentive for behavior. Such accountability does not even specify the "call of duty" needed to reach the desired outcome. This can certainly stifle the perception of personal control needed to develop responsibility. And if the accountability system is viewed as top-down and negative, the probability of people doing anything extra for industrial safety is lessened even more.

So the mission of this book is not only to provide guidelines for developing an accountability system that provides direction and motivation for injury-prevention efforts, but also to teach you how to facilitate personal responsibility for safety. For most of you, I bet the phrase "I'm holding you accountable for that" comes across as negative and perhaps even as a threat. Often such a statement comes after a mistake or an injury, and thus is not proactive. This only indicates lack of personal control. It reflects fault finding, and certainly does not promote personal responsibility. Let's see what it takes to act responsibly for safety. I call the behavior "actively caring."

Actively Caring for Safety

Latané and Darley (1970) studied the helping behaviors of individuals in a variety of emergency situations and proposed that people go through four decision-making steps before choosing to help another individual. As illustrated in Figure 3, we need to notice a problem, believe something should be done about it, assume personal responsibility to do something, and then choose an intervention approach.

Each of these steps in Figure 3 provides a convenient excuse for not helping. "I didn't see the problem!" "I didn't think it was serious enough to require immediate attention!" "Why am I the one who has to deal with this? It's not my responsibility! And besides, I don't know the best way to handle the problem. Someone else could do a better job!" You see, it's easy to talk ourselves out of actively caring.

Acting responsibly for safety is not difficult. Anybody in the workplace can see a hazard or an at-risk behavior, and do something about it. And the more people who believe in going beyond the call of duty for safety, the easier it gets. What's difficult is convincing ourselves and others that it needs to be done. Figure 4 depicts the Safety Triad which I introduced in 1989 as a simple way to characterize a comprehensive approach to improving industrial safety (Geller, Lehman, & Kalsher, 1989). Safety depends upon the three domains shown on that triangle — environment factors, person factors, and behavior factors. Responsible intervention for safety can and should address all three domains.

Three ways to actively care. When you alter a work condition, you are actively caring from an environmental perspective. Posting a warning sign near a hazard, designing or installing a machine guard, and cleaning up a spill you come across are all examples of environment-focused actively caring. They are examples of going beyond the call of duty to protect the safety of yourself and others.

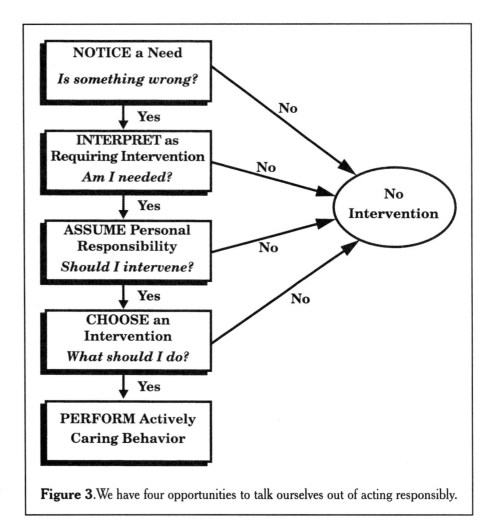

Figure 3. We have four opportunities to talk ourselves out of acting responsibly.

Person-based actively caring occurs when you attempt to make other people feel better. You address their emotions, attitudes, or mood states. This happens when you actively listen to others, inquire with concern about another person's difficulties, or when you send a get-well card to a coworker. Person-based actively caring is likely to boost an individual's self-esteem, optimism, or belonging — which in turn makes it more likely that the person will increase his or her personal responsibility for safety and actively care.

Behavior-based actively caring occurs when you attempt to influence another person's safety-related behavior. Giving recognition for a job well done, coaching a person on a safe work routine, implementing a DO IT process or an incentive program, are all ways to demonstrate personal responsibility for safety with a behavior focus.

Caring directly or indirectly. In each of these domains, actively caring can be direct or indirect. What's the difference? Well, direct actively caring means you are the agent of change. In other words, you fix an environmental hazard; you build a person's self-esteem by asking for advice; or you give a person feedback about a behavior. But even when you don't exert direct influence, you can still actively care indirectly. For example, you can leave a thank-you note for the person on the other shift who cleaned up the work area. Or, you might include a buckle-up reminder at the end of your answering-machine message.

You're the agent of change when you actively care directly.

In the environment domain, indirect actively caring is an easy way to make a difference. All it requires is keeping your eyes open for hazards, and taking personal responsibility for reporting them. It sounds easy, but it doesn't happen often enough.

An illustrative story. Let me tell you about a near catastrophe at an aluminum plant. A guy slipped on an unwelded plate and almost fell several stories to his death. Fortunately, he was able to catch himself with his arms and pull himself back up onto the walkway. The safety committee decided to do more than the typical reactive investigation. They didn't just pin the blame on the welder responsible for securing the plate. They looked for other contributing factors in an effort to prevent similar mishaps. And guess what they discovered? At least a dozen people had slipped on that same loose plate and said nothing about it.

Why didn't people act responsibly in that situation? Well, let's go back to the decision steps that precede actively caring (as depicted in Figure 3). Did those people perceive a problem? Sure they did. The plate slipped when they stepped on it. Did they believe the problem needed immediate attention? Well, maybe not. They didn't fall, so they might have assumed no one else would. Did they think it was their responsibility to get the plate welded? Obviously not. They weren't welders. From their perspective, it was somebody else's job.

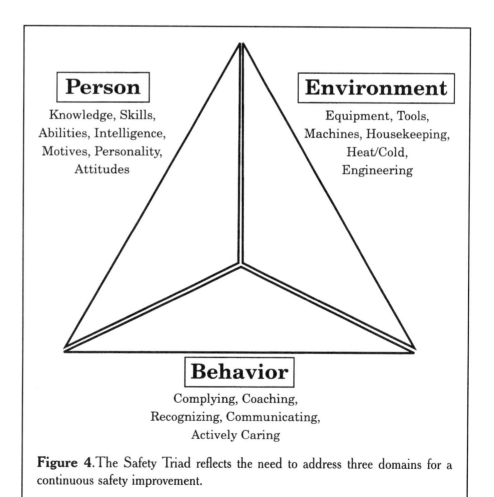

Figure 4. The Safety Triad reflects the need to address three domains for a continuous safety improvement.

> **This is a story about people named Everybody, Somebody, Anybody, and Nobody. There was an important job to be done and Everybody was sure that Somebody would do it. Anybody could have done it, but Nobody did. Somebody got angry about that, because it was Everybody's job. Everybody thought Anybody could do it, but Nobody realized that Everybody would not do it. It ended up that Everybody blamed Somebody when Nobody did what Anybody could have done.**
>
> *Author unknown*
>
> **Figure 5**. Assuming someone else will be accountable can be irresponsible.

Indirect actively caring is easy but important.

"It's not my job" is often the most convenient excuse, isn't it? In other words, if "I'm not held directly accountable for welding a secure metal plate, then it's not my responsibility to intervene when I notice it's loose." But guess what? Any one of those people could have actively cared indirectly and prevented a possible fatality. All they had to do was report that environmental hazard. But as reflected in the words of Figure 5, many people have a convenient excuse for not going beyond the call of duty. They just figure someone else will be responsible.

In Conclusion

Your challenge is to establish the work conditions, procedures, and culture that enable and promote personal responsibility for safety beyond accountability. This is reflected in actively caring behavior which can be direct or indirect and can focus on the environment, person, or behavior side of the Safety Triad. These six types of actively caring behaviors are illustrated in Figure 6.

Direct behavior-based actively caring builds personal responsibility.

Behavior is at the top of the pyramid in Figure 6, signifying that behavior-based actively caring is most challenging and is therefore most infrequent. But people can be held accountable for behavior-based actively caring, and under the right circumstances such action can increase personal responsibility and more actively caring. In this case, involvement precedes commitment. People act themselves into feeling more responsible. This book shows you how to start a spiral of accountability feeding responsibility, feeding actively caring behaviors, feeding more responsibility (and so on), resulting in people becoming totally committed to achieving an injury-free workplace. Start with limiting top-down control, as discussed in the next chapter. You can dictate accountability, but not responsibility. In fact, top-down accountability can decrease responsibility.

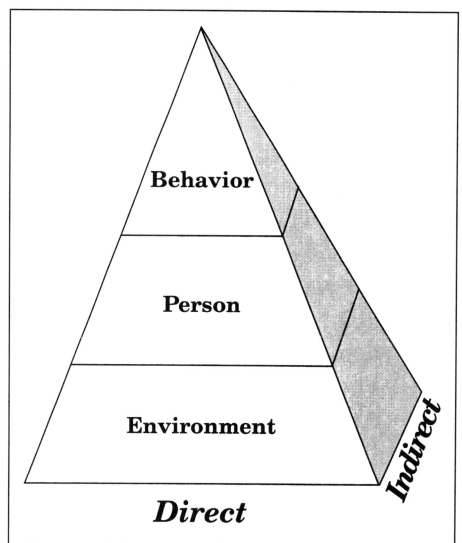

Figure 6. Actively caring can be direct or indirect, and target environment, person, or behavior factors.

Decrease Top-Down Controls

Chapter 3

Occupational safety is too often a confrontation between a rule enforcer and a rule breaker. In other words, safety is viewed by many as compliance with certain safety policies and mandates. In this case, the primary job of safety personnel is to check for worker compliance with safety procedures and to correct (or confront) incidences of noncompliance. Making safety a priority means increasing the enforcement of safety policy. The accountability for this kind of safety usually falls squarely on the plant safety engineers, union health and safety representatives, or perhaps a local safety committee. Occupational safety becomes the unenviable obligation of a select group of individuals. And most often this assignment translates into "monitoring for compliance with safety rules" and "keeping records of injuries" which presumably occurred because of noncompliance. But, as illustrated in Figure 7, a monitoring system is rarely sufficient to hold people accountable for all of their behaviors.

Safety policy implies accountability.

The rigorous enforcement of a particular safety policy, such as consistent power lockout and use of certain personal protective equipment will surely change priorities in favor of "safety." Such a shift in priorities, however, might be quite transitory, and last only as long as the policy is visibly enforced.

What can be done to make compliance with corporate safety policy a standard part of the work routine, accepted and monitored by the workers themselves, and not dependent upon enforcement procedures? In other words, how can we facilitate the adoption of a safety policy into an individual's personal responsibility system of appropriate work performance. Under what conditions will an employee become personally responsible for safe job performance all the time?

What aspects of a corporate culture will facilitate the transfer of safe performance on the job to safe performance at home? What can be done, for example, to increase the probability that workers will not only wear safety glasses, long-sleeved shirts, and steel-toed shoes on the job, but will also wear this protective clothing while they mow their lawns? This reflects the primary theme of this book — How to extend personal responsibility for safety beyond accountability.

Perceptions of Individual Freedom

Answers to the important questions posed above start with a consideration of a basic principle of human nature. People do not like to feel controlled, especially those raised in a culture where individual freedom is a societal imperative worth fighting and dying for. Many psychologists and social scientists have conducted research on the topics of "perceived control" and "perceived freedom," and their conclusions support this principle.

Figure 7. An accountability system can't monitor everything.

Sometimes it feels good to beat the system.

In order to appreciate this aspect of human nature, however, you don't have to read the numerous supportive books and research articles. You need only reflect on your own experiences and perceptions. Recall, for example, your own teenage years. Teenagers undergo a variety of physical and psychological changes, including the notion of "independence" and "individual freedom." As a result, teenagers often break parental rules they had complied with in the past just to assert their own freedom. And this rule-breaking behavior is often followed by reinforcing feelings of independence and personal freedom. It feels good to circumvent authority — to beat the system.

When state laws mandating safety-belt use were initially passed in the mid-1980's, many people complained loudly. It wasn't that they didn't like safety belts, they just didn't want to be told what to do in the privacy of their own vehicles. In fact, this "individual freedom" principle was used by a radio disc jockey in Massachusetts to rally support for successfully rescinding the belt-use law in that state.

After the implementation of state belt-use laws or corporate belt-use policies, several motorists went out of their way to avoid compliance without getting caught. Some beat the system by wearing a T-shirt with a realistic shoulder and lap belt printed on the front. Others purchased and used a strap that could be attached to their shoulder and waist with velcro adhesive in order to give the impression of shoulder-belt use.

Currently, the nationwide belt use rate is about 70%, a substantial increase over the pre-law national belt-use rate of 15%. Thus, many motorists have buckled up to comply with belt-use mandates and for many, safety-belt use has become a healthy habit. At the same time, however, numerous motorists have actively resisted compliance with belt-use laws, and unfortunately these individuals are among the more risky drivers. It will take more than top-down laws to get these individuals to buckle up.

People try to regain personal freedom with countercontrol.

Let me tell you one more real-world story of top-down control influencing bottom-up attempts to assert individual freedom. Psychologists call this "countercontrol." Not too long ago, a top manager at a large oil refining company noticed an operator who was not wearing his hard hat. When he asked the worker about this, the individual touched the top of his head and said, "Oh I didn't realize that, I felt a hat on my head (a baseball cap) and thought it was a hard hat." This manager, figuring he uncovered a "root cause" of noncompliance with hard-hat policy, distributed memos and posted signs to announce a new policy, "No baseball caps allowed." The outcome of this well-intentioned attempt at a quick fix is illustrated in Figure 8. You can't dictate safety!

Figure 8. Top-down control can influence attempts to beat the system.

Instructive Research

Pioneering research by professor Jonathan Freedman (1965) demonstrated the need to limit top-down control if we want people to develop personal responsibility. Dr. Freedman used a mild or severe threat to prevent seven- to nine-year-old boys from playing with an expensive battery-controlled robot.

In the Mild Threat condition, the boys were merely told, "It's wrong to play with the robot." Boys in the Severe Threat condition were told, "It is wrong to play with the robot. If you play with the robot, I'll be very angry and will have to do something about it." Then the experimenter left the room. Four other toys were available for the boys to play with. From a one-way mirror, researchers observed that only one of 22 boys in each condition touched the robot.

About six weeks later, a young woman returned to the boys' school and took them out of class one at a time to perform in a different experiment. She made no reference to the earlier study, but instructed each boy to take a drawing test. While she scored the test, she told each boy he could play with any toy in the room. The same five toys from the previous study, including the robot, were available. Of the boys from the Severe Threat condition, 17 (or 77 percent) played with the robot, compared to only 7 (33 percent) from the Mild Threat condition.

Less control on the outside leads to more control from the inside.

Dr. Freedman theorized that more boys experiencing the mild threat developed an internal rationale for avoiding the robot. As a result, they still would not play with this toy when the external pressure was absent. With less outside control, these children developed internal control over their behavior.

Other researchers have followed up this study to demonstrate that people are more apt to develop internal motivation when external rewards or threats are relatively small and insufficient to completely justify a target behavior. This phenomenon has been referred to as the "less-leads-to-more effect." It's most likely to occur when people feel personally responsible for their choice of action and the resulting consequences. This book explains how to make that happen. First you need to decrease appearances of top-down control, as typified in fault finding.

Eliminate Fault Finding

We do too much fault finding. If we want people to develop personal responsibility for safety, we should be fact finding, not fault finding. But when something bad happens, what do we do? Too often our impulse is to blame people. "Okay let's find out whose fault this was!" Talk about counterproductive. That's not the way to get people involved in the safety process. But we do it all the time. Organizations fault find, supervisors fault find, and on a personal level, we all have a tendency to do it.

Do you have an incident *investigation* team at your plant? What does the term "investigation" imply? Although the *American Heritage Dictionary* (1991) defines the verb "investigate" as "to make a detailed inquiry or systematic examination," the implication of an investigation is to find fault, as in a "criminal investigation." Therefore, I recommend a change in terminology. Why not call this team an incident *analysis* team to communicate a fact-finding mission. In other words, this team's assignment is to study systematically all aspects of the situation in order to find out what can be changed to decease the likelihood of the incident happening again. I offer more details about the operation of an incident analysis team in *Building Successful Safety Teams: Together Everyone Achieves More* (Geller, 1998).

Attribution errors. Let me tell you about the fundamental attribution error. Social psychologists discovered this common human error when they were studying how people explain the behavior of others. What they found is that when we're evaluating the actions of other people, we overestimate the influence of internal factors, and we underestimate the external factors. For example, suppose a coworker is overwhelmed by a production demand. That's the external factor. If he spends the afternoon ignoring you or snapping at you when you talk to him, you're probably not going to judge him in terms of his situation. You're much more likely to blame it on internal factors and say he's rude, or he's just in a really bad mood.

When evaluating others we blame internal factors.

When someone else gets injured on the job, we do the same thing. We're likely to say the person was being careless or sloppy, rather than distracted by a sudden noise. That's how we see things when we're judging others. We tend to look for faults inside the person and use a personal indictment to fix blame.

When evaluating ourselves we blame external factors.

It's a different story, however, when we evaluate our own mistakes. Then we do just the opposite. We don't blame ourselves. We look for external factors to justify our behavior. "Why did I strain my shoulder moving the loaded cart? I'll tell you why. For one thing, Earl on the night shift shouldn't have left the cart where it was. It was his responsibility to move it out of the way. And besides, those carts get too heavy anyway when people overload them."

Figure 9 shows you just how far people will go to avoid taking personal responsibility for a mishap. These are actual quotations from insurance forms submitted by drivers who were involved in automobile crashes. Check out the language people used to keep from admitting any personal blame. Do you see why we should not be fault finding?

The self-serving bias protects our self-esteem.

To protect our self-esteem, we all overestimate external causes and we underplay the internal factors that may have contributed to an error or mistake. Psychologists call this the self-serving bias. And this tendency to protect ourselves from blame becomes very powerful when we feel threatened. If we believe others are looking for ways to pin the blame on us, we're going to hide as much as we can. Who wants to feel like an idiot? Who wants to be criticized for doing something wrong? So as illustrated in Figure 10, finding ways to blame others for problems is much more popular than accepting personal responsibility.

Promote fact finding. If we want to develop people's responsibility for safety, we need to focus on fact finding, not fault finding. An injury or a near miss is usually the result of several causes, and a lot of them may have nothing to do with the person directly involved. Improving safety depends upon getting at *all* causes of an injury. We can't do this if we make people unwilling or frightened to discuss an incident openly, because they're afraid of being blamed, or of receiving a penalty. Removing fear of failure is a critical part of developing internal feelings of personal responsibility.

- **The other car collided with mine without giving warning of its intentions.**
- **A pedestrian hit me and went under my car.**
- **The guy was all over the road. I had to swerve a number of times before I hit him.**
- **I had been shopping for plants all day, and was on my way home. As I reached an intersection, a hedge sprang up, obscuring my vision. I did not see the other car.**
- **As I approached the intersection, a stop sign suddenly appeared in a place where no stop sign had ever appeared before. I was unable to stop in time to avoid the accident.**
- **An invisible car came out of nowhere, struck my vehicle, and vanished.**
- **My car was legally parked as it backed into the other vehicle.**
- **The pedestrian had no idea which direction to go, so I ran over him.**
- **The telephone pole was approaching fast. I was attempting to swerve out of its path when it struck my front end.**

Figure 9. People are reluctant to admit personal blame for their vehicle crashes (excerpted from *The Toronto Sun*, 1977).

Figure 10. It's popular to find fault in others.

The reporting of a near-miss should be rewarded.

We need continuous help in improving safety, and the people out there doing the jobs are the best help we've got. Instead of making them hesitate to report a near miss, we should be rewarding them for bringing it to our attention. You can turn any near miss into an opportunity to achieve something for safety. The worker involved should be an active participant in a systematic analysis of the incident. With his or her assistance, a corrective action should be implemented to reduce the likelihood of someone else suffering the same experience. This is the kind of involvement that can build personal responsibility for safety.

At one company I visited, near misses and the improvements they led to were reported in a monthly newsletter, and the employee who completed the near-miss report was thanked for helping the safety effort. At another company, workers gave videotaped accounts of incidents. The worker involved described how the near miss occurred, and then explained what changes were made to help prevent a similar situation. Copies of the video were then distributed to other plants in the company. The division manager viewed the videotapes, and then wrote a letter of commendation to the worker who experienced and reported the near miss.

Instead of being blamed, these people were given the chance to present the facts and educate the work force. Would punishing these workers have done as much to improve their safety performance? Would it have helped the safety process nearly as much? Of course not. Instead of feeling like losers, they achieved a small win for safety. Let's turn to a more complete discussion of punishment — its misuses and liabilities.

The Errors of Punishment

For too many people, holding people accountable means punishing them for the mistakes they make. In industry we call this "discipline." This is unfortunate because one Latin root of discipline — disciplina — means instruction or training, and another — discipulus — refers to a learner. Research shows punishment to be an ineffective teaching tool, and in most cases it does more harm than good.

I've met many managers who include a "discipline session" as part of the corrective action for an accident. Injured employees receive a lecture from a manager whose own safety record and personal performance appraisal were tarnished by the injury. These "discipline sessions" are unpleasant for both parties, and certainly do not encourage personal responsibility or commitment to the safety mission of the company.

Punishment stifles personal responsibility.

Instead, the criticized and embarrassed employees are simply reminded of the top-down control aspects of corporate safety, usually resulting in increased commitment not to volunteer for safety programs, or to encourage others to participate. In this case, the culture loses the involvement of invaluable safety participants. Individuals who have been injured on the job have special insight into conditions and behaviors that can lead to an injury. If these individuals can be persuaded to discuss their injuries with others, they can be very influential in motivating safe work practices. Personal testimonies (especially by people known to the audience) have much greater impact than statistics summarizing the outcomes of a remote group.

In other sources, I have covered the various undesirable side effects of punishment (Geller, 1996a, 1997a,c), and discussed bottom-up ways to implement punishment when it's absolutely necessary (Geller, 1998). Here it's most appropriate to consider the type of work behaviors that results in injury — human errors and calculated risks. In order to reduce the possibility of property damage or personal injury, people need to be held accountable for their errors and risky behaviors. But this doesn't mean they should be punished for performing them. In fact, I hope to convince you that human error never warrants punishment, and only rarely does a calculated risk deserve disciplinary action.

The Information Processing Cycle

First, it's useful to understand where errors, mistakes, and calculated risks come from. What aspects of our daily functioning are responsible for at-risk behavior, whether intentional or unintentional? If we know where to look for the causes of at-risk behaviors, we can do something to reduce them.

Consider that we are continually taking in, processing, and reacting to information. Almost everything we do results from this basic information processing cycle. We sense a stimulus (input); we evaluate the stimulus and plan a course of action (interpretation and decision-making); and then we execute a response (output).

All errors are unintentional.

Input stage. All errors are unintentional and are most likely at the input and output stages of the information processing cycle. When we fail to notice a certain stimulus or misread a dial, we are at risk for an input error. Sometimes our own physical limitations can restrict our ability to detect, recognize, or identify a critical stimulus or environmental event. At other times, changes in environmental conditions (like a glare on a dial) can cause input errors.

Often our own prior experiences can influence these errors, as when we tune out familiar and seemingly insignificant stimuli. This error-causing process is called habituation, and occurs naturally when we experience stimuli with no relevant consequence. Habituation helps us attend to important stimuli, especially in an information-overload situation. But habituation can be detrimental to safety when a previously unimportant stimulus suddenly becomes relevant because of a change in consequences.

Mistakes are unintentional.

Interpretation and decision-making stage. This component of information processing is the judgment phase, and is therefore the location of most mistakes and calculated risks. In this stage we interpret our sensory input and decide on a course of action. But our judgment might be faulty and lead to risky behavior. For example, consider you give someone explicit instructions on how to perform a particular task safely. Let's assume the individual hears everything you say, but still doesn't use all of the personal protective equipment (PPE) you requested. If this person did not understand your instructions and did not think the PPE was needed when it was, an unintentional at-risk behavior or mistake was performed. You might consider this "unconscious incompetence." In contrast, intentional risk taking occurs if the individual understood your instructions but didn't want to take the time to put on PPE which was thought to be unnecessary. This is considered a calculated risk or "conscious incompetence."

Calculated risks are intentional.

In both cases, the at-risk behavior resulted from poor judgment or decision making. Maybe the individual knew the PPE was required but just forgot to put it on. Perhaps the person didn't appreciate the seriousness of your PPE request. Maybe he thought "it won't happen to me," and thus chose to take the short cut. Or it's possible the individual noticed a conflict or inconsistency between your request and the behavior of others (since some other workers were not using the PPE), or between your request and the leadership of a supervisor (who seems to value productivity over safety). Perhaps the person perceived the PPE as uncomfortable, silly looking, or inconvenient, and looked for an excuse to ignore your advice.

Obviously, a number of factors can guide this stage of information processing toward poor judgment and an unintentional mistake or calculated risk. Through frank and open communication with this person, you could pinpoint factors influencing the at-risk behavior, and make appropriate corrective adjustments. This requires the cultivation of a special kind of work culture. One in which people willingly hold themselves accountable for their at-risk behaviors — whether intentional or unintentional — and look for ways to decrease the possibility of similar mistakes or calculated risks.

Output errors can occur when you are not fit for work.

Output stage. This is the execution phase of information processing and corresponds to the safe or at-risk behavior we can observe objectively. An error occurs at this stage when the individual is unprepared to perform the required task. This can be due to limitations in strength, reaction time, coordination, or endurance. And such limitations can be due to a variety of psychological or physical variables, including fatigue, distress, alcohol, age, arthritis, or a cumulative trauma disorder, to name a few.

This type of error often occurs when a work setting does not adequately account for the various characteristics and limitations of all humans. Operators can usually adjust to these situations, but occasions arise when the mismatch between machine and a particular human results in an output error.

Slips vs. Mistakes

As I've already explained, our mistakes and calculated risks occur after perception and before execution — when we evaluate the situation and decide on a course of action. Given that mistakes result from incorrect intention or judgment, this makes perfect sense. But what about slips, the input and output errors of our ways?

To understand the cognitive source of a slip, let's consider the various types of these errors, as defined by Donald A. Norman in his classic book, *The Psychology of Everyday Things* (1988). I highly recommend this book for those who want to learn more about human errors and ways to reduce them.

Capture errors. Have you ever started traveling in one direction (like to the store) but suddenly find yourself on a more familiar route (like on the way to work)? How many times have you borrowed someone's pen to write a note or sign a form, and later found the pen in your own pocket? Professor Norman calls these "capture errors" because a familiar activity or routine seemingly "captures" you and takes over an unfamiliar activity. This error seems to occur at the execution stage of information processing. However, it also involves misperception or inattention to relevant stimuli, as well as the absence of conscious judgment or decision making.

Capture errors can be caused by old habits.

How does this error "slip" into the work routine? Have you ever started a new task and found yourself performing behaviors relevant for another more familiar routine? Has a change in PPE requirements influenced any slips? It seems reasonable that a routine way of doing something (even at home) could "capture" your execution of a new work process, and lead to a slip and an injury.

Safe habits reduce capture errors.

This is one reason to get in the habit of practicing the safe way of doing something, regardless of the situation. Then your safe behavior is put in automatic mode, and slips can actually be advantageous. This happens when you reach for the shoulder belt in the back seat of a vehicle because of your habitual buckle-up behavior as a driver. And when you use your vehicle turn signal at every turn, this safe behavior also becomes habitual. Then there's no need to think about this safe and courteous driving behavior. In this case the "capture" is beneficial to your safety. When basic safety-related behaviors become habits, we have more mental capability for higher-level thinking, and the probability of a mistake is reduced.

Description errors. These slips occur when the description or location of the correct (safe) and incorrect (at-risk) executions are similar. For example, I have thrown a tissue in our clothes hamper instead of the waste can, even though our clothes hamper is not next to a trash receptacle. On a few occasions, I've actually thrown dirty clothes in the trash can, and once I threw a sweaty T-shirt in the toilet. According to Professor Norman, the similar characteristic of these three items — a large oval opening — led to these slips.

Similar control switches can cause description errors.

Do you have any switches on operating equipment that are similar and in close proximity, but control different functions? How unsafe is it to throw the wrong switch? Many control panels are designed with this error in mind. Switches or knobs controlling incompatible functions are not located next to one another, and they often look and feel distinctly different for quick visual and tactile discrimination. It might be useful to identify similar-yet-different behaviors in your workplace that lend themselves to description errors.

Loss-of-activation errors. Have you ever walked into a room to do something or to get something, but when you got there you forgot what you were there for? You think hard but just can't remember. Then you go back to the first room you were in and suddenly you remember what you wanted to do or get in the other room. Why do these things happen?

Forgetting can lead to a "brain cramp" or a loss-of-activation error.

This slip is not rare, and is commonly referred to as "forgetting." Dr. Norman refers to it as "loss-of-activation" because the cue or activator that got the behavior started was lost or forgotten. This happens whenever you start an activity with a clear goal, but then lose sight of it as you continue the task. You might continue the task, but with little awareness of the rationale for progress toward a goal. With regard to the three stages of information processing I introduced above, this error starts in Stage 1 — input but eventually affects the output stage when you can't complete the task without more information. And the interpretation and decision-making stage is involved because this is where lapses in memory occur.

What are the implications of this kind of slip for industrial safety? First, recall the basic ABC contingency of behavior-based safety — "A" for activator, "B" for behavior, and "C" for consequence. Since this error results from the loss of activation for a task, we can prevent this kind of slip by adding specific behavior-based activators to various situations. Through signs, goal-setting discussions, verbal reminders, and friendly hand gestures we attempt to instruct or guide performance. Figure 11 suggests we need friendly reminders or activators.

When people tell you they already know what to do with statements like, "Stop harping on the same old thing," you can say you're just actively caring by trying to prevent a "loss-of-activation" error. You'll never know how many of these slips you'll prevent. But you can motivate yourself to keep activating by reflecting on your own experiences with this sort of "brain cramp" and then considering the large number of people in your work setting who have similar slips everyday.

Mode errors. This error is probable whenever we face a task involving multiple options, or modes of operation. These slips are inevitable when equipment is designed to have a greater variety of functions than the number of control switches available. In other words, when controls are designed for more than one purpose, depending on the mode of operation, you can expect occurrences of this error.

Mode errors occur when one switch controls multiple task options.

Over the years, I've owned a variety of digital watches with a stopwatch mode. Each one has had a different arrangement of switches designed to provide more functions than control buttons. Therefore, the meaning of a button press depends upon the position of a mode switch. So, guess how many times I've pressed the wrong button and illuminated the dial or reset the digital readout when I only wanted to stop the timer? Have you experienced the same kind of mode error, if not with a stop watch perhaps with the text editor of a personal computer? Airline pilots must be especially wary of this kind of error.

Figure 11. Activators can be helpful in preventing cognitive failures or memory lapses.

This type of slip is essentially one of execution. But these errors often occur because we forget the mode we're in. This involves memory and the interpretation and decision-making phase of information processing. Equipment design can certainly reduce the frequency of this error, but so can proper training and behavior-based activators. When we work with people whose job allows for mode errors, we need to be alert to the behaviors and conditions that can lead to this slip and then conduct actively caring observation and feedback.

The Role of Experience

It's important to understand the difference between slips, mistakes, and calculated risks. And we need to realize when one type of at-risk behavior is more probable than others. For example, with more experience and perceived proficiency on the job comes the greater likelihood for a slip. That is, as people become more competent and confident, they pay less deliberate and conscious attention to what they are doing. They automatically filter out certain stimulus inputs, they do less interpretation and decision making, and they resort to automatic modes of execution.

Mistakes and calculated risks are possible among both beginning and experienced workers. New hires make safety-related mistakes when they don't know the safe way to perform a task or when they don't understand the need for the special safety precautions. They take calculated risks when the activators or consequences they receive from others favor the at-risk alternatives.

Experienced workers make mistakes when they take safety for granted and fail to consider the injury potential of a certain at-risk behavior. They take calculated risks when they feel especially skilled at a task, realize they have never been seriously injured at work, and consider the soon and certain benefits of an at-risk behavior to outweigh the improbable costs.

Taking Responsibility for Our Errors

Conversations about personal errors and calculated risks are key to injury prevention.

Understanding the variety of potential slips, mistakes and calculated risks in the workplace should lead to a realization that most of these go unnoticed or are ignored. In other words, when our at-risk behaviors don't lead to personal injury, we just forget them or explain them away. In fact, this is basic human nature. Who likes to talk about their errors?

We feel much better talking about our good times than our bad times. But I'm sure you recognize injury prevention requires a shift in perspective. Only through open and frequent discussion about our safety-related errors and calculated risks can we alter the environmental conditions that can reduce them. By revealing our own slips and mistakes in group meetings, we raise awareness and motivation to be accountable for at-risk behavior. Then with a proper behavior-based observation and feedback system, it's possible to anticipate and prevent the errors of our ways, and to prevent personal injury. Tell yourself that when you see at-risk behavior capable of causing personal injury, you will be taking a serious calculated risk if you don't take appropriate action to stop it.

Figure 12 summarizes five types of at-risk behaviors as being unconscious (unintended) or conscious (intended) incompetence. The list introduces the term "violation." Whenever punishment or so called "discipline" is involved, it's necessary to consider whether a specific safety rule was intentionally violated. As discussed above, slips and mistakes are unintentional, and while they might be clearly unsafe and violate a safety rule, it seems unfair to apply punishment. In both situations, the person meant well. The at-risk behavior occurred because of inaccurate information processing.

If a safety rule covers the situation and at-risk behavior, it can be violated intentionally or unintentionally. If a person didn't know the rule or forgot the rule, the at-risk behavior should be considered an "unintended violation." This is unconscious incompetence. In contrast, a calculated risk is an "intended violation." It occurs when the violator is aware of the safety rule, but for a number of possible reasons, chooses to ignore it. Personal accountability for this type of at-risk behavior often includes a punishment contingency.

Punishment and Accountability

When is punishment a viable approach to reducing at-risk behavior? I propose you consider seven basic questions before applying a punishment contingency. Answering these questions will help you decide whether a person deserves a negative consequence for performing an unsafe or at-risk behavior. In most cases, you'll find that another type of corrective action is more appropriate. Some answers will offer direction as to the type of corrective action that might facilitate more responsibility for the unsafe act you're holding the person accountable for.

> 1) **Slip** *(unconscious incompetence)*
>   - **Behavior is unintentional or unplanned, and results from a cognitive failure or "brain cramp."**
>
> 2) **Mistake** *(unconscious incompetence)*
>   - **Behavior results from poor judgment due to a deficit in knowledge, skill, or experience.**
>
> 3) **Calculated Risk/No Violation** *(conscious incompetence)*
>   - **Behavior is at-risk but no safety rule covers the situation.**
>
> 4) **Calculated Risk/Unintended Violation** *(conscious incompetence)*
>   - **Behavior violated a safety rule the person does not know about or forgot.**
>
> 5) **Calculated Risk/Intended Violation** *(conscious incompetence)*
>   - **Behavior violates a safety rule the person knows about and remembers.**
>
> **Figure 12.** Five types of at-risk behavior can account for injury.

1. Was a Specific Rule or Regulation Violated?

You can't have a policy for every possible at-risk behavior.

If you answer "no" to this question, punishment is obviously unfair. Does this mean we need to write more rules or document more regulations? I don't think so. You can't write a rule for every possible unsafe behavior. Yet, proper injury analysis will reveal some human error to be more at risk for causing serious injury than behaviors already covered by a rule or regulation. But since human errors are unintentional, rules won't decrease them. We need to allow for the possibility that non-compliance with an existing rule or regulation can be unintended. This leads to the next question.

2. Was the At-Risk Behavior Intentional?

All human error is unintentional. We mean well but have cognitive failures or "brain cramps." As I discussed earlier, psychologists call these slips or lapses, and they are typically due to limitations of attention, memory, or information processing.

Have you ever walked into a room and forgot why you were there? Have you ever locked yourself out of your car with the keys still inside? Have you ever forgot where you left your car in a multi-level parking complex? Have you ever left the house more than once in the morning, returning each time to get something your forgot? How many times have you missed an exit or turn on a familiar road, or found yourself en route to a familiar but unintended destination?

Unconscious incompetence. In the workplace, research has shown these types of errors increase with an individual's experience on the job. Skilled people put their actions on "automatic pilot," and perhaps add at-risk behaviors to the situation. How many of us fiddle with a cassette tape or juggle a cellular phone while driving? Under these circumstances, an error can easily occur.

Mistakes are caused by unintentional poor judgement.

With mistakes, the individual was well-intentional regarding the ultimate outcome of getting the job done, but used poor judgment in getting there. Have you ever miscalculated your gap in a parking area and scraped an adjacent vehicle? How many times have you planned a poor travel route and therefore got caught in traffic congestion you could have avoided? Have you ever pressed the brake pedal too quickly on a slippery road or pumped the brakes in an antilock system? These are all examples of mistakes—at-risk behaviors performed unintentionally and inappropriately under the circumstances. Usually we don't give these errors much thought until they lead to a vehicle crash. Such unconscious incompetence needs to be corrected, but certainly not through punishment.

Calculated risks are caused by intentional poor judgment.

Conscious incompetence. Sometimes poor judgment is used to intentionally take a risk. As I mentioned earlier, this is a calculated risk. Suppose you don't buckle your safety belt. Or perhaps you divide your attention between the road and some other task like map reading, phone dialing, or cassette selecting. You know these behaviors are unsafe or risky, but you decide to take a calculated risk? In this case, unlike a mistake, you are aware that your behavior is inappropriate. Such behavior might seem rational because it's not followed by a negative consequence and is supported with perceived comfort, convenience, or efficiency.

The deliberate or willful aspect of calculated risks might seem to warrant punishment, but punishment won't convince people their judgment was defective. And that's what is needed to change conscious incompetence into conscious competence.

3. Did the Person Know and Remember the Violated Rule?

Mistakes decrease with experience.

Calculated risks increase with experience.

Researchers have found an inverse relationship between a person's experience on the job and the probability of injury from a mistake. In other words, the more knowledge or skill we have at doing something, the less likely we are to be unconsciously incompetent. So the driving mistakes listed above are performed more often by inexperienced or poorly trained drivers. On the other hand, the tendency to take a calculated risk increases with experience on the job. This is human nature, and it won't be changed with punishment.

This question reflects the fact that some at-risk behaviors occur because the rule or proper safe behavior was not known. It's possible for an experienced worker to forget or inadvertently overlook the rule that made the at-risk behavior a violation and a target for punishment. Training and behavior-based observation and feedback can reduce these types of errors, but punishment certainly won't help.

4. How Much Were Other Employees Endangered?

Intentional risk taking doesn't imply intentional endangerment.

This question reveals the rationale I hear most often for punishing at-risk behavior. Many safety professionals claim they only use punishment in their workplace for the most serious matters. And this is when a particular behavior puts the individual at risk for a severe injury or fatality, or places many individuals in danger. Failing to lock-out a power source during equipment adjustment or repair is the behavior most often targeted for punishment. Some are quick to add, however, that for punishment to be warranted, this behavior must be made "knowingly and willfully."

"Knowingly" means the individual knew a rule was violated (Question 3), and "willfully" refers to the intentional issue (Question 2). Now if an employee willingly and knowingly avoided a lock-out procedure to put himself and others at risk for injury, then the severest punishment is relevant. Actually, this person should probably be fired immediately. But this rationale for at-risk behavior is very rare.

Many at-risk behaviors result from poor judgment. But whether they are unconscious (mistakes) or conscious (calculated risks), most are not committed with the idea that someone will get hurt. Often specific characteristics of the work environment or culture enable or even encourage a calculated risk. Thus, the next question needs careful consideration.

5. What Supports the At-Risk Behavior?

This is the most important question of all. People don't make errors or take calculated risks in a vacuum. Poor judgment occurs for a reason. And it's important to learn an employee's rationale for taking a risk. This can lead to useful corrective action.

Find out a person's rationale for at-risk behavior.

Did the individual lack appropriate knowledge or skills? Was a demanding supervisor or peer pressure involved? Did equipment design invite error with poorly labeled controls? Was the "safe way" inconvenient, uncomfortable, or cumbersome? Was it viewed as inefficient and unnecessary?

Let's look at the organizational culture. Is safety taken seriously only after an injury? Is safety performance judged only in terms of injuries reported per month, instead of the number of preventive activities?

These are only some of the questions that need to be asked when finding out what factors must be changed in order to reduce the possibility of an error or calculated risk reoccurring. Punishment will only stifle the kinds of conversations needed to answer these questions and develop a practical corrective action plan.

6. How Frequent is the At-Risk Behavior?

Numerous at-risk behaviors precede injury.

A particular error or calculated risk is investigated and punishment is considered because something called attention to its occurrence. In other words, the accountability process is probably a reaction to an injury. How many at-risk behaviors typically occur before leading to an injury? H. W. Heinrich (1931) estimated 300 near misses for every one major injury, and Frank Bird (1969) observed this ratio to be 600 to one (as cited in Bird & Germain, 1997). Both Heinrich and Bird presumed numerous at-risk behaviors occur before even a near miss is experienced, let alone an injury.

So what good is it to punish one of many risky behaviors? If the behavior is an error, punishment will only stifle reporting and the search for potential remedies. If the probability of getting caught while taking a calculated risk is low — and it's minuscule if you wait until an injury occurs — the threat of being punished will have little behavioral impact. However, the threat of punishment can influence the kind of attitude reflected in Figure 13. This can decrease responsibility for safety.

Punishment does much more to inhibit involvement in safety improvement efforts than it does to reduce at-risk behavior. So consider your observation of an at-risk behavior a mere sample of many similar at-risk behaviors, and use the occasion to stimulate interpersonal dialogue about ways to reduce the behavior's occurrence.

Figure 13. The threat of punishment can lead to decreased responsibility.

7. How Often Have Others Escaped Punishment?

Inconsistent punishment reduces trust.

One sure way to lose credibility and turn a person against your safety efforts is to punish a person for an at-risk behavior others have performed without receiving similar punishment. To the individual, it's irrelevant that he or she had performed the behavior in the past without your negative attention. The fact is the person has seen others perform similar behaviors without punishment, but now he or she is getting picked on. This is viewed as unfair. Perceived inconsistency is the root of mistrust and lowered credibility. So why risk such undesirable impact, especially when beneficial behavioral influence is improbable anyway?

The Futility of Punishment

By discussing the various causes of at-risk behavior, I hoped to show the futility of using punishment as a corrective measure. Errors (cognitive failures and mistakes) are unintentional and are often caused by environmental or cultural factors. And when an error is intentional (as in a calculated risk), the person didn't intend to cause an injury. Rather there were factors in the situation that influenced the decision to take the calculated risk. The factors contributing to errors and calculated risks need to be discovered and addressed.

The threat of punishment for at-risk behavior stifles open conversation about preventing at-risk behavior.

If the threat of personal injury is not sufficient to motivate consistent safe behavior, it doesn't help to add one more threat to the situation. In addition, the threat of punishment for at-risk behavior can impede responsibility for reporting errors and calculated risks, and looking for ways to reduce them. This is opposite of what you want in a proactive approach to industrial safety and health promotion. We need open and frank conversations with the people working at analyzing and changing management practices, equipment, or organizational systems that contribute to much of the at-risk behavior we see in the workplace.

The other day I was thinking about the various types of at-risk behavior during a tennis match. I always make a number of errors and calculated risks on the tennis court. Reducing these errors is key to winning. Errors occur, for example, when my focus or concentration wavers and I unintentionally "miss-hit" the ball, or fail to get in position for a solid return. Calculated risks occur when I rush the net at the wrong time or hit the ball long when trying to hit a baseline corner.

The only way to reduce my tennis errors on-the-spot is to quickly refocus my attention or reevaluate the situation. What I often do instead is engage in a self-defeating punishment strategy. I yell at myself internally, and sometimes even talk out loud. Once in a while a run of errors influences me to hit the ball against the fence or even to throw down my racket. Does this self-critical "punishment" ever help? Of course not. It only makes matters worse. The same is true for your golf game, and for meeting the continuous challenge of preventing injuries in the workplace.

Context and Accountability

On a ski weekend last February I was reminded of the dramatic influence "context" has on at-risk behavior. According to my *American Heritage Dictionary* (1991), context refers to "the circumstances in which a particular event occurs." It includes both the outside and inside stuff surrounding people when they are performing. This refers to what we see others doing on the outside, and how we feel on the inside — from feelings of competence, confidence, and commitment to perceptions of insecurity, uncertainty, and risk.

I use a simple demonstration to illustrate the influence of context for my students in introductory psychology. I ask volunteers to simultaneously stick one hand in a bucket of ice water and the other in a bucket of hot water (around 100°F). After about 10 seconds, I ask the volunteers to remove their hands from the two buckets and put both hands in a third bucket filled with water at room temperature (about 70°F). However, they do not experience room temperature. In fact, with one hand the water appears quite warm, while with their other hand the water seems rather cool.

Perceived experience is influenced by the context in which behavior occurs.

You don't have to be there to appreciate how the prior brief temperature exposure influenced subsequent perception. In fact, you've probably already guessed which hand experienced warm water and which hand experienced cold water. We live this simple context effect everyday. Coming indoors from the cold gives the impression of warmth, but in contrast to a hot summer day our home can appear quite cool. Yet my students express surprise when they experience "warm" in one hand and "cool" in the other while soaking both in the same bucket of water.

So how does this simple demonstration of contextual contrast relate to my skiing experience? More importantly, what does all of this have to do with industrial safety and health? Let me start by explaining the circumstances surrounding my ski trip. This was only the third time in my life I had ever tried to ski. The first time was in 1974. Furthermore, the hills were quite icy, and the so-called beginner hills appeared to be quite steep. I didn't see a "bunny hill" anywhere, but my daughter urged me on. So even with low competence and confidence and perceptions of uncertainty and high risk, I took to the crowded slopes. One near miss after another did not stop me, and nor did one "wipe out" after another.

My numerous bruises qualified for several OSHA recordables. My only consolation was that I was not the only one in pain. The next day, many guests at the lodge were limping around. Some were sitting with legs wrapped and elevated — more OSHA recordables. Most other skiers in my age range were much more experienced than I, yet several told me they were having a difficult time because of the icy conditions. Their admonitions were not sufficient for me to ignore my daughter's urgings, "Come on dad, just one more hill; you can do it." I was also influenced by the "big bucks" I had paid for this ski weekend. I wanted to get my money's worth.

Risky behaviors in one context generalize to other situations.

The risky behavior of the slopes generalized to the ski lifts. And here lies the real context lesson of my story. The lift chairs had protective restraining bars that could be pulled down conveniently. The signs requesting the use of these "restraining bars" hardly seemed necessary. The need for this protective device was obvious.

The lifts rose to heights over 300 feet above the ground. It wouldn't take much for someone to slip off the seat, especially given the slick material of most ski pants. And when the lift stopped, the chairs rocked forward and backward slightly, making the need for this protective device even more evident. But here's the kicker — the bottom line. More often than not, I observed the bars in the upright position. Most skiers were not using this protective device. Did the risky context of the skiing experience influence decreased use of this protective device?

At every lift, a "courtesy patrol" person guided lines of people to the entrance, and another individual helped people take their seats. There was ample opportunity for these "professionals" to remind skiers to use the protection device. However, I never heard such a reminder. And the many long lines I stood in that weekend gave me numerous opportunities to hear such a safety message.

CHAPTER 3 — DECREASE TOP-DOWN CONTROLS

Within the personal context of inexperience or reduced confidence, we hesitate to take control.

My daughter's friends rode the lift several times without pulling down the protective bar because they didn't know it was there. I, however, noticed the protective device, and used it every time — well, almost every time. I must confess that once my daughter and I rode a lift with two young men who appeared to be expert skiers. This time I didn't pull down the bar, at least not at first. Instead, I waited for one of "the experts" to take control. Within the context of my insecurity and reduced self-confidence, I waited for someone else to intervene. Only when our chair stopped and rocked a bit, about 200 feet above the ground, did I reach up to pull down the safety bar. There I was, a researcher and educator who has studied and lectured about safety for over 25 years, hesitating to protect my daughter, myself, and two strangers.

Context is my only excuse for such irresponsible behavior. Not only did the use of this protective bar seem insignificant within the context of the greater perceived risk of skiing, but I hesitated to take control within the context of two experienced skiers. I might add that the two "experts" seemed quite perturbed at my eventual protective behavior. They both grimaced slightly, with one having to move his ski poles to make room for the protective bar. And long before we reached the end of our ride, one "expert" raised the protective bar, presumably preparing to dismount.

Obviously the environmental context at this busy ski resort was not conducive to proactive protective behavior. Must it take someone falling off a chair-lift before attention will be given to promoting the use of these fall protection devices? Does the risky context of a ski resort make it more difficult to get people involved in proactive safety initiatives?

Context at Work

Like any workplace, a ski resort is a mini-culture, with its own set of rules, norms, behavioral patterns, and attitudes. And all of this was influenced by the overriding purpose or mission of the resort — to give people the exhilarating experience of gliding down snow-covered hills. Nowhere in the resort's mission statement was there a message about safety. Actually, for some people, attempting to link safety with skiing might seem inconsistent. Afterall, skiers spend their money to take extraordinary risks. Why should we look out for their personal safety?

A top-down context stifles employee involvement.

Does the mission statement of your industry reflect an overarching concern for production and quality? Is safety considered a priority (instead of a value) that gets shifted when production quotas are emphasized? Is safety viewed as a top-down condition of employment rather than an employee-driven process supported by management? Are safety programs handed down to employees with directives to "implement per instructions" rather than "customize for your work area?" Does the top-down perspective from OSHA, as illustrated in Figure 14, influence a top-down safety context at your plant?

Are safety initiatives discussed as short-term "flavor-of-the-month" programs rather than an ongoing process that needs to be continuously improved to remain evergreen? Are near-miss and injury investigations perceived as fault-finding searches for a single cause rather than fact-finding opportunities to learn what else can be done to reduce the probability of personal injury? Are the elements of a safety campaign or injury analysis considered piecemeal factors independent of other organizational functions rather than aspects of an organizational system of interdependent functions?

Figure 14. The government's top-down approach to safety can be detrimental to the work context.

The traditional context of safety management stifles employee involvement.

Are employees held accountable for outcome numbers that hold little direction for proactive change and personal control rather than process numbers that are diagnostic regarding the achievement of an injury-free workplace? Do employees take a dependency stance toward industrial safety whereby they depend on the organization to protect them with rules, regulations, engineering safeguards and personal protective equipment?

A "yes" answer to any of these questions implies a barrier in your work context that needs to be overcome in order to achieve the ultimate injury-free workplace. Getting employees involved in safety is difficult within the context of top-down rules, regulations, and programs supported almost exclusively with the threat of negative consequences. In contrast, employee involvement and responsibility for safety is much more likely with top-down support of safety processes developed, owned, and continuously improved upon by work teams educated to understand relevant rationale and principles. This is the kind of context in which actively caring can be cultivated and a Total Safety Culture eventually achieved.

Outcome numbers do not provide a diagnostic context for injury prevention.

Metrics used to evaluate the safety performance of individuals, teams, and the organization as a whole have a powerful influence on context. Employee responsibility, ownership, and involvement can increase or decrease depending on the accountability system employed. Injury statistics, for example, provide an overall estimate of distance from a vision of "injury-free," but they are not a diagnostic tool for proactive planning. If used as the only index of safety achievement (or failure), injury-related outcome numbers can do more harm than good, alienating people rather than empowering them to take control of safety. On the other hand, numbers that measure the quantity and quality of process activities related to safety performance provide the accountability context needed to motivate individual and team responsibility. They direct continuous improvement of the process. Clearly this is not rocket science. It is merely basic principles of total quality management applied to occupational safety and health. Developing the kind of accountability system that can increase workers' empowerment or personal responsibility for safety is covered in the next chapter of this book.

In Conclusion

If we want people to go beyond the call of duty to protect themselves and others from personal injury, we need to cultivate the kind of culture or work context that promotes personal responsibility for safety. This calls for paradigm shifts in management practices, accountability systems, and interpersonal relationships. This chapter focused on the need to decrease the common perspective that industrial safety is a top-down control process meant to hold employees accountable for compliance with certain rules dictated by government regulation or corporate policy.

Threats of punishment inhibit the development of personal responsibility.

The paradigm of top-down control is promoted with slogans like "safety is a condition of employment" and supported with threats of negative consequences for rule infractions. Under this traditional top-down paradigm, enforcement is the key to motivating human behavior, and is actually one of the three E-words of traditional safety management (the others being engineering and education). So after the engineers develop safe machinery and personal protective equipment, people need to be educated on safe work practices. And when people don't follow the rules or the prescribed operating procedures, enforcement is used to motivate compliance. This means punishment, or as they say in industry — "discipline."

The top-down enforcement approach will motivate compliance, but that's all. Within an enforcement context, don't expect people to feel responsible for safety or do more than they can be held accountable for with rules and regulations. Basically, the more control people feel on the outside, the less control they will develop on the inside. In other words, holding people accountable with top-down threats of punishment will not facilitate personal responsibility. In fact, if people think their work context reduces their individual freedom, they might attempt to regain it by looking for ways to by-pass rules or beat the system.

Personality responsibility for safety develops in a context of genuine concern for injury prevention.

People are more likely to develop personal responsibility for safety in a work culture that shows genuine concern for reducing personal injury. Such actively caring for safety is demonstrated when fact finding is promoted over fault finding, and when a proactive approach is taken toward incident analysis. In this case, proactive means that near misses, property damage, and at-risk behaviors are analyzed to find ways of reducing them, thus preventing injuries. This type of analysis will rarely include punishment, because it's understood that negative consequences will only stifle the openness and interpersonal trust needed to conduct a useful incident analysis and derive a practical and effective corrective action plan.

Increase Feelings of Empowerment for Safety

Chapter 4

In the management literature, empowerment refers to delegating authority, sharing decision making, or holding people accountable to achieve something. In this book, I'm not talking about this typical definition of empowerment, which can be viewed as a top-down strategy for getting people to do more with less. Instead, I'm talking about the psychological side of empowerment. When I say empowered I mean feeling empowered. How do you feel about receiving more empowerment — about being held accountable for more? It's not about receiving more responsibility; it's about feeling more responsible. It's about believing you can make a difference with the added "power" and feeling good about that. In this context, feeling empowered for safety means the same as feeling responsible for safety.

Empowerment is in the eyes of the beholder.

Figure 15 illustrates the distinction between giving empowerment (or responsibility) versus feeling empowered (or responsible). The manager genuinely believes the added assignment is worth a congratulation, but the recipient of the additional duty is not pleased. He doesn't feel more responsible.

There could be several reasons for not feeling empowered in this situation. The recipient might not believe this new position is important or meaningful. Perhaps he doesn't feel capable of handling the accountability, either because he lacks the necessary knowledge, skills or resources, or because

Figure 15. Empowerment is in the eyes of the beholder.

he thinks factors outside his personal control will influence his effectiveness at the task. Or it's possible he just doesn't expect very many positive outcomes to result from his new assignment. He's not optimistic about his new leadership role.

These possible reactions to a task assignment reflect three different person states which determine perceptions of empowerment or responsibility — self-efficacy, personal control, and optimism. When you increase any of these feeling states in yourself or others, perceptions of responsibility are increased. The opposite occurs when these states are lowered. So by understanding these person states and what factors influence them, we can determine how to facilitate empowerment or personal responsibility.

> *Personal responsibility is determined by feelings of self-efficacy, personal control, and optimism.*

I have discussed these empowerment states at length in other sources (Geller, 1996a, 1997a,c). Here I only review their meaning briefly before discussing safety-related conditions that can influence them and personal responsibility for safety. First, self-efficacy means simply the belief that "you can do it." You believe you have the education, the training, the resources, and the time to do what needs to be done.

When you have personal control you believe you're in control. You expect your own efforts to determine outcomes — not luck, not chance, and certainly not the arbitrary decisions of others. So your personal control is increased whenever you see a cause and effect relationship between your efforts and measures of accountability. When people are held accountable for numbers they don't believe they can directly control, their personal control is sapped. This can lead to decreased empowerment or personal responsibility.

It's important to realize that personal control is in the eyes of the beholder. You might think certain conditions should increase a person's perceptions of control when, in fact, they don't. It's also true that many people can have more control over situations than they think. Often it's a matter of perspective. Although most people consider the experimenter in control of the situation illustrated in Figure 16, the mice have another idea. And, their perspective is not wrong. The term "top-down" is actually misleading. It can influence people to overlook the reciprocal influence between everyone working interdependently in a system, regardless of job title.

Now, what about optimism? You are optimistic when you expect to be effective or successful at what you do. When people expect the best they are likely to get the best. That's the self-fulfilling prophecy. Optimistic people expect their efforts to pay off. And that anticipated payoff motivates them to keep going and keep meeting challenges. Figure 17 illustrates the distinction between an optimist and a pessimist, who doesn't expect the best and therefore is less motivated to make a difference.

When people feel empowered and responsible, they have the passion to go beyond the call of duty for the safety of themselves and others. When their internal script says "I can do it," "I'm in control," or "I expect the best," they get involved in the safety process — not because they have to, but because they want to. And when we give them the methods and tools to make a difference, they'll use them. They'll act on their caring. An empowered workforce feels responsible for safety. They believe they're in control of safety and can make a difference. They actively care. Let's consider the type of accountability system that will help make this happen.

Figure 16. In an interdependent system everyone has control.

Hold People Accountable for Numbers They Can Control

The most accurate pop psychology slogan I've heard goes like this, "What gets measured gets done, and what gets measured and rewarded gets done well." A related slogan is "Feedback is the breakfast of champions." You've probably heard these or similar words before. They are key to continuous improvement.

It's intuitive that we need feedback to improve. Therefore it's obvious that continuous improvement depends upon having the right kind of evaluation process or accountability system in place. People's performance goals are based on what they're held accountable for, so their effort to achieve their goals also depends on what's being measured. The type of performance goals set determines whether people feel personally responsible for achieving the goals. This in turn influences performance. So the accountability system for safety determines the extent to which people go beyond the call of duty for safety. In other words, your accountability system influences the amount of actively caring occurring in your work culture.

Outcome-Based Accountability

How is safety performance measured at your facility? Does your company do like most and focus on outcome numbers like the number of injuries that occur? Does your safety department gauge their improvement by the TRIR or the total recordable injury rate? Do people at all levels of your organization believe they're personally accountable for the TRIR?

Figure 17. Pessimists and optimists have different perspectives.

For many organizations the TRIR represents their ultimate safety score — their "true" measure of safety success. Many companies, in fact, use this index to rank their plants with regard to safety performance. Often the TRIR is used to determine which organization wins an annual safety award. This is the primary statistic used by federal and state regulatory agencies to evaluate whether an industry needs to be audited with regard to their compliance with various safety standards.

Holding people accountable for their plant's TRIR reduces personal responsibility for safety.

It's no wonder company employees, even line workers, are held accountable for the plant's TRIR. At least that's how it feels when TRIR or "the number of days worked without a lost time injury" is the primary or only safety statistic displayed or discussed at safety meetings. This can make an employee feel accountable not only for his injuries but also for the injuries of others, including injuries to employees outside his immediate work group. As such, people can feel they're being held accountable for a number of injury-causing factors outside their personal control, such as damaged equipment, environmental hazards, excessive workloads, system contingencies causing a hurried work pace, and a culture that promotes an attitude of "rugged individualism" rather than group cohesion or an interdependency perspective. You can bet that such an accountability system will only stifle or decrease personal responsibility for safety.

A subjective indicator. As illustrated in Figure 18, many factors unrelated to occupational health and safety influence the TRIR. Some of these are certainly outside the control or domain of responsibility of the individual employee. Although the TRIR is often displayed as an accurate and reliable statistical "fact" based on objective, indisputable data, this is not the case. Actually, the factors contributing to the TRIR can be quite subjective and unreliable. Beyond the obvious causes of injury — at-risk behavior and environmental hazards — consider the following contributors: a) the willingness of an employee to report a personal injury; b) the decisions and influence an injured employee's supervisor can have on whether an injury gets reported; and c) the diagnosis and decision of the treating physician, who can cause an injury to be classified as "recordable" simply by the type and amount of drugs or treatment he or she prescribes.

TRIR is a reactive measure of failure.

Consider also that a) TRIR is a measure of safety *failures* rather than safety *successes*, and as such focuses on *reactive* correction, rather than *proactive* prevention; b) it's probably inappropriate to compare one plant's TRIR to another's, especially if the plants have work processes implicating different levels of risk; and c) the varying criteria defining an OSHA recordable makes it inappropriate to compare a TRIR of today with a TRIR of 5 or 10 years ago — even at the same plant.

Inconsistent reporting. Suppose an employee at your organization bends down to pick up a pen while walking up a flight of stairs, and suddenly his back "goes out." He is taken to the hospital and diagnosed with a ruptured disc in the lower back, resulting in several weeks of lost work days for therapy and recuperation. Would that injury be counted as an OSHA recordable and added to the lost workday statistic at your plant? Although most safety professionals would answer "yes" to both of these questions, there would be exceptions. In fact, at one plant where this actually happened, the back injury was not counted as an OSHA recordable and thus did not tarnish the plant's exemplary TRIR.

The plant safety director merely followed the company's policy of calling the back injury a "life event." Just because the debilitating injury was experienced on company property doesn't mean the corporation should be held accountable. The injury was the cumulative result of repeated shock to the lower back from activities performed in many

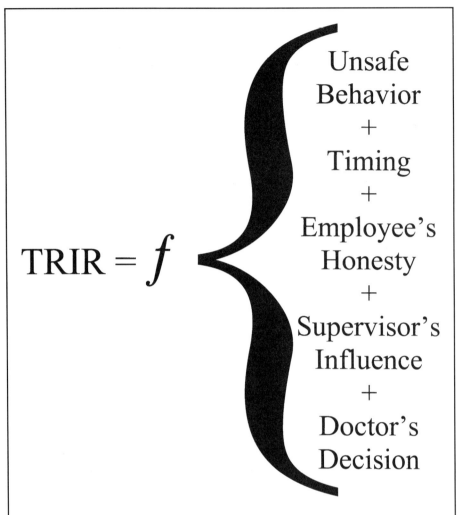

Figure 18. Many factors unrelated to safety influence the total recordable injury rate (TRIR).

locations, including the individual's home. This is the rationale used to discount this and similar incidents as an OSHA recordable. Once OSHA fined this company $18,000 for not counting one of these "life events" as an OSHA recordable. The fine was paid but the company continues to maintain its reasonable policy of not counting "life events" in its TRIR.

Focusing on the TRIR can sap personal responsibility for safety.

My point here is that you have every right to question the validity of the outcome statistics used to a) evaluate corporations' safety performance, b) rank plants within a company according to safety accountability, and c) assign bonus checks to plant managers and safety directors. More importantly, a safety accountability system based on failures and perceived as uncontrollable saps people's responsibility for safety. How can workers feel motivated to improve safety when the only feedback they receive is an announcement that someone got injured? Outcome statistics put people's focus on the wrong thing. It's like trying to play baseball with your eye on the scoreboard instead of the ball.

Losing perceived control. The detrimental impact of this kind of accountability system was shown dramatically at two large chemical facilities where I consulted several years ago. Figure 19 shows how the 1800 employees of a large industrial complex were held accountable for safety. The man on the ladder (twice life size and named "I. M. Ready") climbed one step higher each day there was no lost-time injury. Whenever a lost-time injury occurred, I. M. Ready fell down the ladder and started his climb again. In addition, a red light at the entrance/exit gate flashed for 12 hours after a lost-time injury. Every employee was promised a reward as soon as I. M. Ready reached the top of the ladder to signify 30 days without a lost-time injury.

At first this plan activated significant awareness, even enthusiasm, for safety. But no specific tools or methods were added to reduce the injury rate. Safety did not improve,

Figure 19. Holding employees accountable for zero injuries can have undesirable side effects.

and I. M. Ready did not reach the top of the ladder in two and a half years. Initial zeal for the program waned steadily. Eventually, people stopped looking at the display. The man on the ladder and the flashing red light were reminders of failure. Most of the employees did not feel personally responsible for the failure, but they didn't know what to do to stop the injuries. Many workers became convinced they were not in direct control of safety at their facility and developed a sense of learned helplessness (Seligman, 1975) about preventing lost-time injuries.

Process-Based Accountability

There is a happy ending to my story of an outcome-based accountability system that stifled employee responsibility for safety. The outcome-based incentive program was dropped, and a process-based approach was implemented. An incentive/reward program was implemented which held employees accountable for doing things for safety. The behavior-based accountability system, developed by an incentives/rewards team (as described in Geller, 1998), was essentially a "credit economy." Various safety-related behaviors, achievable by all employees, earned a certain number of "credits." At the end of the year, participants exchanged their credits for a choice of different prizes, all containing a special logo.

The variety of safety-related behaviors earning credits included: attending monthly safety meetings; preparing materials for a safety presentation; leading a safety meeting; writing, reviewing, and revising a job safety analysis; and conducting periodic audits of equipment, environmental conditions, and designated work practices. For a work team to receive credits for their audit activities, the results of its environmental and PPE observations had to be posted in the relevant work areas.

It's noteworthy that only one behavior was penalized by a loss of credits. This was the result of the late reporting of an injury. Employees were reminded that a minor injury left untreated can lead to an OSHA recordable. The plant nurse can usually detect when this happens. Thus, this safety incentive program held workers accountable for reporting all injuries immediately, in order to receive early treatment and avoid more serious complications (such as an infection).

A behavior-based incentive program holds people accountable for things they can control.

At the start of the new year, each participant received a "safety credit card" for tallying ongoing credit earnings. Some individual behaviors earned credits for the person's entire work group, thus promoting group cohesion and teamwork. The audit aspects of this incentive/ reward program exemplify an essential process for preventing occupational injuries — interpersonal observation and feedback.

I'm not suggesting you need a behavior-based incentive/reward program to promote the kind of safety accountability that can increase feelings of personal control and responsibility, but I am saying you need an accountability system that focuses on safety-related process activities people can control. Top management needs to keep worrying about the injury numbers. But don't hang this accountability on the people doing the jobs on the floor. Let these people focus on the process — the day-to-day behaviors and activities going on in the workplace. They can control and feel responsible for these things.

When workers receive feedback that lets them know they are reducing factors that can contribute to an injury, they are more apt to care about making the process work. They own the process because they believe it helps them control their own destinies. When people feel empowered to do what's needed to reduce injuries in the workplace, they accept the responsibility of making safety work. If someone gets hurt, they take it personally. Then they look for process things they can do to prevent the injury from happening again. Setting the right process goals is critical for establishing accountabilities for which people feel responsible.

Safety goal-setting. Included among Dr. W. Edwards Deming's fourteen points for quality transformation are "eliminate management by objectives, eliminate management by numbers, numerical goals . . . eliminate slogans, exhortations, and targets for the work force asking for zero defects" (Deming, 1985, p. 10). Does this mean we should stop setting safety objectives and goals? Should we stop trying to activate safe behaviors with signs, slogans, and goal statements? Does this mean we should stop counting OSHA recordables and lost-time cases, and stop holding people accountable for their work injuries?

Goal-setting is effective when done correctly.

Answers to all of these questions are "yes," if you take Deming's points literally. However, my evaluation of Dr. Deming's scholarship and workshop presentations, along with my personal communications with Dr. Deming in 1990 and 1991, have led me to believe that Deming meant we should eliminate "goal-setting, slogans, and work targets" as they are currently implemented. Dr. Deming was not condemning the principles of goal-setting, management by objectives, and activators; rather he was criticizing the current corporate use of these principles. Substantial research supports the use of objective goals and activators to improve behaviors, if these behavior change principles are applied correctly. Setting zero injuries as a safety goal and exhorting this goal or objective in corporate newsletters and mission statements is misuse of these principles and should in fact be eliminated.

Setting goals incorrectly. Holding people accountable for numbers they do not believe they can control is a sure way to produce negative stress or distress. For some people such incorrect goal-setting will not result in distress because they won't take such goals seriously. Their experience has convinced them they cannot control the numbers, so they simply ignore the goal-setting exhortations. As I mentioned above, these individuals have overcome the distress of unrealistic management objectives or goals by developing a helplessness perspective or attitude.

What does the goal of zero injuries mean anyway? Is this goal reached when no work injuries are recorded for a day, a month, six months, or a year? Does a work injury indicate failure to reach the goal for a month, six months, or a year? Does the average worker believe he or she can influence goal attainment, beyond avoiding personal injury?

Effective goals are specific, motivational, achievable, recordable, and trackable.

Set SMART goals. Every safety book I've written has included a section about the importance of setting SMART goals. That's because the research literature is filled with success stories following goal-setting when the goals are specific with regard to measurable actions or behaviors, and when progress in achieving the goals is tracked objectively as feedback. Feedback is the consequence that maintains motivation for continued goal-directed behavior.

I remember the techniques for setting effective goals with the word SMART — "S" for specific, "M" for motivational, "A" for achievable, "R" for recordable, and "T" for trackable. In my book on team building (Geller, 1998), I added the letter "S" for shared, because team members hold each other mutually accountable for achieving the various aspects of a team goal.

Chapter 4 — Increase Feelings of Empowerment For Safety

Set goals with the end in mind.

Motivational means that consequences are specified in goal-setting. That is, the effective goal-setter defines the consequences available when the goal is reached, and tracks the accomplishment of steps toward achieving the goal. The feedback from completing intermediate steps toward ultimate goal achievement serves to motivate continued progress. Of course, it's critical that the people asked to work toward goal achievement "buy" or own the goal, and believe they have the skills and resources needed to achieve it.

Safety goals should focus on process activities that can contribute to injury prevention. Employees need to discuss the variety of things they can do to reduce workplace injuries, from reporting and investigating near misses to conducting safety audits of environmental conditions and work practices.

Set goals on process activities.

Practical examples. A safety steering team I worked with a few years ago wanted to increase daily interpersonal communications regarding safety. They set a goal for their group to achieve 500 safety communications within the following month. To do this they had to develop a way to measure and track safety communications. They designed a wallet-sized "SMART Card" for recording conversations with persons about safety. One member of the group volunteered to tally and graph the daily card totals.

Another work group I consulted with set a goal of 300 behavioral observations of lifting. In other words, the employees agreed to observe each other's lifting behaviors according to a critical behavior checklist they had developed. If each worker completed an average of one lifting observation per day, the group would reach their goal within the month. Each of these work groups reached their safety goals within the expected time period, and as a result they celebrated their "small win" at a group meeting—one with pizza and another with jelly-filled donuts.

These two examples illustrate the application of SMART goals. They depict safety as process-focused and achievement-oriented, rather than the traditional and less effective outcome-focused and failure-oriented approach exemplified by injury-based goals. These examples also illustrate an employee-driven approach to safety rather than the more typical and less effective top-down paradigm. The workers were motivated to initiate a safety accountability process because it was their idea — they owned the process. And, they stayed motivated to work on their safety process, because through SMART goal-setting they knew where they were going, they knew when they got there, and they tracked their progress along the way.

Vision versus goal. Figure 20 shows a simple flow chart summarizing the basic approach to safety accountability from a behavior-based perspective. You start a safety improvement mission with a vision or ultimate purpose. This could be "zero injuries" or an injury-free workplace. Then, with group consensus for the vision, you develop methods, procedures, or action plans to accomplish your mission. These are the process-oriented goals which individuals or teams are accountable for achieving. Each goal, of course, implicates certain behaviors needed to achieve it. Since these behaviors are specified with SMART goals, holding people accountable to achieve specific goals means they are accountable to perform designated behaviors.

Start with a mission statement then set relevant process goals.

The most effective accountability systems provide extrinsic consequences for behaviors that support SMART or SMARTS goals. Such consequences might be formal as in a) a behavior-based incentive/reward program (like the "credit economy" I discussed earlier), or b) a systematic tracking system for recording safety conversations or obser-

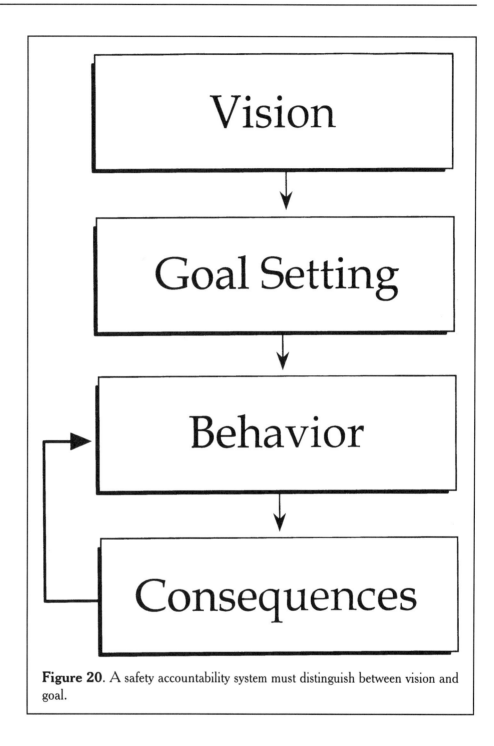

Figure 20. A safety accountability system must distinguish between vision and goal.

vations of lifting behaviors (as discussed above). Or, the behavioral consequences might be informal, as with one-on-one delivery of praise or recognition. In either case, the consequences are both informative and motivating. They tell the recipient he or she is being accountable, and they motivate the individual to keep performing the desired behavior. Positive consequences are also likely to influence people to extend their personal responsibility beyond explicit accountabilities.

Reinforce goal attainment with recognition and rewards.

A common line among "pop psychologists" is that "vision" is more motivating than specific goals. I'm sure you've heard the story that people who look beyond their daily tasks to a larger purpose or mission add meaning to the daily humdrum of a typical workday. This seems quite intuitive. I bet all of you can reflect on times when your attitude toward a job was more positive when you visualized the larger enterprise or system to which you were contributing.

Figure 21 illustrates the distinction between goal and purpose as presented by many pop psychologists. The worker on the left sees himself as merely a brick layer and therefore is presumably less motivated than the worker who perceives his job as contributing to a team with a larger mission. Similarly, an employee who appreciates the connection between the completion of daily safety-related behaviors and the vision of an injury-free workplace will probably feel more personal responsibility for safety than the worker who puts up with everyday safety inconveniences in order to comply with OSHA rules and regulations. However, we shouldn't lose perspective here and disregard the value of setting specific behavioral goals.

People with purpose set challenging goals.

You should assume that the happier and more motivated worker in Figure 21 has accomplished more work because he has set and achieved more challenging goals. In other words, the greater purpose for his job motivated him to set SMART goals and to work hard to achieve them. His vision or destination did not detract from focusing on the

Figure 21. Goals provide accountability and vision increases motivation.

Figure 22. Appreciation of the process can be lost by over-focus on the end result.

process or journey. And his team members offer interpersonal appreciation for each other's daily goal-directed contributions to the overall mission. They appreciate the fact that "success is a journey not a destination."

I hope you see my point in dwelling on the distinction between vision and goal. The vision or purpose of "zero injuries" is a motivating destination, but we can only get there by setting process goals relevant to the journey. Unfortunately, this basic principle is too often ignored or abused, especially in safety. I've seen the vision of an injury-free workplace get translated into the following:

- a nonspecific, useless goal of "zero injuries" which gives no direction or opportunity to celebrate,

- slogans like "safety is a condition of employment" and "all injuries are preventable" that seem top-down and restrictive rather than empowering,

- outcome-based incentive programs that put peer pressure on individuals not to report an injury, and

- punishment contingencies or so called "discipline" programs that seeming treat employees like children requiring adult supervision instead of the safety authorities they are for their work areas and the responsible safety leaders they need to be.

Success is a journey, not a destination.

Holding people accountable for only results — the injury statistics — instead of the process reminds me of the way many people eat popcorn. Does the illustration in Figure 22 remind you of anyone you know? Have you ever seen people approach popcorn in that way? With popcorn flying everywhere and mouths stuffed to the hilt, the focus seems to be on the outcome of getting filled up with popcorn, instead of enjoying the consumption process. Actually, it's not farfetched to consider such rapid consumption of popcorn to be at-risk behavior. Have you ever choked on a piece of popcorn or other food particle because of an overly rapid eating pace?

Can you imagine any intrinsic benefits to slowing down the consumption of popcorn to one kernel at a time? Besides decreasing the risk of choking, you might actually appreciate this food more. You could savor the distinctive taste of each piece and look forward to the next. And eating at a slower rate would likely result in lower caloric intake. Are you saying to yourself, "so what?" or wondering what all this popcorn stuff has to do with safety?

I can think of a number of real-world situations analogous to the popcorn eating illustration. How often do we rush through daily tasks with little regard or respect for the process. Thinking only of the outcome at the end of the job, we work at a faster-than-safe pace. Perhaps the anticipated outcome is only to get the job done, so we can get to the break room. But often, as with popcorn eating, slowing down a bit can be more enjoyable as well as safer.

Slow down to appreciate the process.

Perhaps you can't relate the popcorn story to a work setting. However, there is one situation relevant to all of us — driving. Imagine the benefits of putting more focus on the means rather than the ends while driving. Over emphasis on the destination rather than the journey is undoubtedly a root cause of numerous vehicle crashes as well as discourteous driving and that horrible U.S. epidemic of the '90's — road rage.

There are certainly consequences intrinsic or inherent to many tasks that can reinforce a focus on the process. However, it's often necessary to use external prompts and rewards to get initial participation. For example, slowing down driving speed offers rewards, or the natural process, of lower gasoline use and greater opportunity to enjoy roadway scenery or in-vehicle conversation. However, extra activators, or consequences, from vehicle occupants might be necessary to overpower the apparent rewards offered by traveling at excessive speeds. The same is true for decreasing a work pace at risk for a cumulative trauma disorder. The principle of shaping is relevant here because it guides intervention that can not only make behavior more accountable for safety, but can also increase personal responsibility.

Principle of Shaping

Behavior-based safety focuses on holding people accountable for process activities they can control. And, positive reinforcement or recognition is used as often as possible so people feel good about their accountability and develop personal responsibility. In this regard, it's important to understand the method of behavioral shaping.

Shaping procedures were developed more than 40 years ago in animal research laboratories, and used extensively by behavior analysts to improve human performance. Before describing the key aspects of shaping, let me tell you how I demonstrated this process 30 years ago to a large class of college students taking introductory psychology. At the time, I was a graduate research assistant at Southern Illinois University.

One day, I intentionally arrived to class before the instructor, hopped up on the stage, and announced to the students that we were going to "shape" the professor, Dr. Neil Carrier, to walk off the stage. I told them it would be easy if they paid attention and listened intently to my quick instructions. They were "all ears" as I quickly said the following:

> Here's what we do. Whenever the professor steps closer to the end of the stage, show undivided attention by looking at his face and smiling or nodding your head to indicate sincere approval and appreciation. When the instructor steps back from a position close to the stage, give your normal reaction to his lectures. Take this opportunity, for example, to write notes or check your other notes, if appropriate. Please, do not cause any disruption at these times. Act normal. The important thing is to give more than usual attention whenever Dr. Carrier gives a closer approximation to our target behavior — walking off the stage.

Perhaps you're thinking that I was not very safety conscious in those good old days. You're absolutely right. Fortunately, Dr. Carrier did not step off the stage, but he came darn close. Within 15 minutes, the professor was delivering his lecture at the very end of the stage, and each of the more than 400 students in the classroom had their eyes glued on him, with some smiling and others nodding their head to show understanding or approval. The teacher was shaped to perform an at-risk behavior, apparently without his awareness.

After maintaining the target behavior for about three minutes, Dr. Carrier stepped back from the edge of the stage and noted an obvious change in the students' behavior. Then, he stepped up to the edge and noted the increased attention the students gave him. He stepped back, and again observed decreased eye contact from many students. Then with a smile, he shouted, "Geller, what did you do?" "Nothing, sir," I replied from the back of the room. Pointing to the students, I continued, "It was your students. They used basic shaping techniques to influence your behavior without your knowledge. Wasn't that a great demonstration?" He could not disagree.

Shaping rewards successive approximations to a target behavior.

Although I certainly don't recommend this technique to get professors or other people to perform unsafe acts, the key aspects of the shaping process reflect pivotal aspects of applied behavior analysis directly relevant to improving safety. Specifically, shaping consists of the strategic use of positive consequences to increase the rate that a desirable behavior occurs. The focus is on observable and desirable behavior rather than undesirable behavior and unobservable attitudes. Plus, positive rather than negative consequences are used in the management process. Note also that the person receives a positive consequence for closer and closer approximations to the target behavior. You can't shape by waiting for perfection. Instead, you observe closely for improvement and support it with reward or recognition procedures.

Understanding Behavior-Based Safety: Step-by-Step Methods to Improve Your Workplace (Geller, 1997c) offers specific guidelines for delivering positive consequences in a shaping process. Here I want you to notice the key aspects of shaping that make it a primary process for increasing both accountability and responsibility for safety. Shaping means targeting the behavior you want, observing carefully for successive approximations to the target, and using positive consequences to reward improvement.

Develop a Comprehensive Accountability System

In many respects accountability is synonymous with evaluation. In other words, the scheme used to evaluate safety performance defines accountabilities. It tells people what they're accountable for. And as I've reiterated several times in this chapter, using injury statistics to evaluate safety performance is not only insufficient, it can do more harm than good when diagnostic process measures are not available. When evaluation focuses on numbers people can control, their empowerment increases and they work harder to impact the bottom line. Thus, whether people develop accountability for safety and accept personal responsibility depends on how you evaluate safety in your organization.

I have documented ways to develop a comprehensive evaluation system for safety in other sources (Geller, 1996a, 1997c). Therefore, I won't go into much detail here, except to offer some key points to consider. Actually, one of the key strengths of behavior-based safety, especially the DO IT process reviewed in Chapter 1, is that it provides objective methods for evaluating and tracking the progress of upstream injury prevention efforts. This approach does exactly what the title of this chapter calls for. It holds workers accountable for numbers they can control. Let's see how the behavior-based paradigm fits into a broader safety evaluation scheme.

Figure 23 summarizes the relationship between various types of upstream measures of safety and the bottom line — injury reduction. It integrates various evaluation domains used to derive safety accountabilities. For example, you could take a knowledge test to check your accountability to learn something useful at a professional development seminar, or you could complete an environmental or behavioral audit to determine whether desirable change has occurred.

Behavior can be viewed as a process or an outcome.

The lower levels of the hierarchy represent process activities needed to improve the higher-level outcomes of a safer environment and ultimate injury reduction. The immediate causes of injury reduction are changes in environment or behavior (or both). Therefore, behavior can actually be viewed as an outcome, in that changes in knowledge, perceptions, or attitudes are pursued in order to affect changes in behavior. Of course, intervention processes can be implemented to change behavior directly so as to reduce the probability of injury.

Figure 23 points out the relativity of process and outcome. An education process, for example, can led to behavior change (an outcome), while a process to change behaviors might result in outcome changes to the environment (as in honed tools and a tidier and organized work area). And, completing a work process in a safer environment can affect the ultimate outcome — a reduction in injuries.

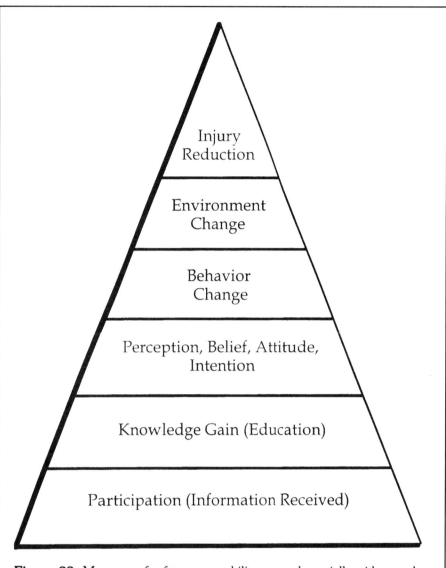

Figure 23. Measures of safety accountability vary substantially with regard to direct influence on injury.

Figure 23 also reflects the three basic areas requiring attention for injury prevention — environment, behavior, and person. This is, of course, the Safety Triad I introduced earlier (Figure 4) to classify different types of actively caring behaviors. The hierarchical levels of intervention impact in Figure 23 reflect one or more of these three domains and suggest a particular approach to evaluation or accountability. While changes in environments and behaviors can be assessed directly through systematic observation, changes in knowledge, perceptions, beliefs, attitudes, and intentions are only accessible indirectly through survey techniques, usually questionnaires. Let's review these three basic approaches to evaluation.

What to Measure?

Most safety intervention programs focus on either environmental conditions — including equipment technology — or human conditions, as reflected in employees' perceptions, attitudes, and behaviors. It might seem reasonable to evaluate change in only the particular area we've targeted for change — environment, behavior, or person state. Consequently, when the target of change is corporate "culture," employee perceptions or attitudes are typically evaluated. If behavior change is the focus of a safety intervention, then certain behaviors are observed and analyzed in terms of their frequency, rate, duration, or percentage of occurrence. And when environments or engineering technologies are evaluated, measurements are taken of mechanical, electrical, chemical, or structural characteristics.

Assessing human factors requires attention to both the behavior and person domains.

I've heard culture-change consultants advocate the substitution of perception surveys for environmental audits. And presentations on behavior-based safety emphasize direct observations of work practices, often in lieu of the subjective evaluation of personal perceptions and attitudes. But given the need for employees to "feel good" about a behavior-based safety process and develop a sense of responsibility beyond accountability, I think it's obvious we need to check people's perceptions and attitudes and ongoing behaviors.

An evaluation of a simple change in equipment design would often benefit from an assessment of the relevant human factors like employees' work behaviors around the new equipment and their attitudes and perceptions about the equipment change. I hope you can see that a comprehensive evaluation process and accountability system for safety requires a three-way audit process to cover a) environmental conditions, b) safety-related behaviors, and c) person states such as perceptions, attitudes, beliefs, and intentions.

A comprehensive accountability system tracks behavior, person, and environment.

Evaluating environmental conditions. In many ways, environmental audits are the easiest and most acceptable type of evaluation. In fact, regular environmental or housekeeping audits are already standard practices at most companies. Most of these evaluations can be improved by involving more employees in designing audit forms, conducting systematic and regular assessments, and posting the results in relevant work areas.

The behavior-based safety incentive program I described earlier in this chapter awarded employees weekly "credits" for accomplishing these components of environmental evaluation. The company recognized the need to involve as many employees as possible in the regular auditing of environmental conditions, including tools, equipment, and operating conditions. This was a positive way to hold people accountable for maintaining safe environmental conditions, and it helped to cultivate their personal responsibility for safety.

Elsewhere I've presented environmental checklists for safety that can be used to graph for public display the percentage of safe conditions and the percentage of potential corrective actions taken for at-risk tools, equipment, or operating conditions (Geller, 1996a, 1997c). My associates at Safety Performance Solutions teach work teams the rationale behind the environmental checklist, and then assist them in applying the checklist in their plant. Employees then customize the checklist and graphing procedures for their particular work areas. Regular audits and feedback sessions increase accountability for environmental factors that can be changed to prevent an injury.

Critical behavioral checklists are generic or job-specific.

Evaluating work practices. The systematic auditing of work practices is the foundation of behavior-based safety, as summarized in Chapter 1. There I explained the DO IT process as an all-inclusive approach to holding work teams accountable for increasing safe behaviors and decreasing at-risk behaviors. For more details on this approach please refer to *The Psychology of Safety* (Geller, 1996a) or *Understanding Behavior-Based Safety* (Geller, 1997c). There I explain how to develop two types of observation checklists: a) a generic version applicable anywhere to assess basic work practices, such as prescribed lifting techniques and use of certain personal protective equipment; and b) a job-specific checklist for particular tasks, like the safe driving checklist my students and I developed to hold drivers accountable for safe vehicle operation.

Behavioral observations can be conducted on groups or individuals.

My experience has been that group auditing — of people using or not using personal protective equipment, for example — is readily accepted by most employees and relatively easy to implement. But one-on-one behavioral audits, whereby employees observe other employees who volunteer to be monitored, are not readily accepted in some corporate cultures. A plant-wide education and training intervention is usually necessary to teach the rationale and procedures for this evaluation process and to develop the necessary interpersonal understanding, empathy, and trust. I detail methods for increasing interpersonal trust in *Building Successful Safety Teams* (1998), and review some of these later in Chapter 6.

Evaluating person factors. Person factors refer to subjective or internal aspects of people. They are reflected in commonly used terms like attitude, perception, feeling, intention, value, intelligence, cognitive style, and personality trait. You can find many surveys that measure specific person factors of target populations ranging from children to adults. Some of these factors are presumed to be *traits*, others are considered *states*. It's important to understand the difference when you consider the evaluation potential of a particular survey.

Person traits are permanent characteristics.

Person *traits* are relatively permanent characteristics of people. They don't vary much over time or across situations. Since traits are relatively permanent, questionnaires that measure them cannot gauge the impact or progress of a culture-change intervention. Trait measures serve as a tool to teach individual differences, but in safety management their application is limited to selecting people for certain job assignments. Obviously, you can't hold people accountable for changing their person traits.

Person states are temporary characteristics.

In contrast, person states can change from moment to moment, depending on situations and personal interactions. When our goals are thwarted, for example, we can be in a state of frustration. When experiences lead us to believe we have little control over events around us, we can be in a state of apathy or helplessness. Person states can influence behaviors. Frustration, for example, often provokes aggressive behavior, and perceptions of helplessness inhibit constructive behavior and facilitate inactivity.

Now I hope you recognize a primary theme of this book. Responsibility is a person state that influences desirable behaviors beyond those monitored by an accountability system. As detailed in this book, various methods and changes in conditions can enhance a person's state of responsibility. This in turn increases one's willingness to go beyond the call of duty for safety and health improvement.

Responsibility for safety is a person state.

Measures of person states can be used to evaluate perceptions of culture change, and to pinpoint areas of a culture that need special intervention attention. Like most culture surveys, the Safety Culture Survey used by my associates at Safety Performance Solutions asks participants to rate statements on a five-point continuum (from highly disagree to highly agree) about their perceptions of the safety culture. Issues include the perceived amount of management support for safety, the willingness of employees to correct at-risk situations and look out for the safety of coworkers, the perceived risk level of the participant's job, and the type of interpersonal consequences following an injury.

Chapter 4 — Increase Feelings of Empowerment for Safety

Our survey also measures factors that increase one's willingness to accept responsibility for another person's safety. These include self-esteem, belonging, and empowerment. Sample items from our survey that measure the actively caring person states are given in Figure 24. They were adapted from professional measures of these characteristics, and have been evaluated for reliability and validity. I've also presented these items elsewhere (Geller, 1996a, 1997c), as well as items from the safety perception and attitude portion of our survey. There's nothing special about the items in this part of our culture survey. They merely ask employees to react to straightforward statements about safety management and improvement.

This is a questionnaire about your beliefs and feelings. Read each statement, then circle the number that best describes your current feelings. There are no "right" or "wrong" answers. This questionnaire only asks about your personal opinions.

	Highly Disagree	Disagree	Not Sure	Agree	Highly Agree
1. I feel I have a number of good qualities.	1	2	3	4	5
2. Most People I know can do a better job than I can.	1	2	3	4	5
3. On the whole, I am satisfied with myself.	1	2	3	4	5
4. I feel I don't have much to be proud of.	1	2	3	4	5
5. When I make plans, I am certain I can make them work.	1	2	3	4	5
6. I give up on things before completing them.	1	2	3	4	5
7. I avoid challenges.	1	2	3	4	5
8. Failure just makes me try harder.	1	2	3	4	5
9. People who never get injured are just plain lucky.	1	2	3	4	5
10. People's injuries result from their own carelessness.	1	2	3	4	5
11. I am directly responsible for my own safety.	1	2	3	4	5
12. Wishing can make good things happen.	1	2	3	4	5
13. I hardly ever expect things to go my way.	1	2	3	4	5
14. If anything can go wrong for me, it probably will.	1	2	3	4	5
15. I always look on the bright side of things.	1	2	3	4	5
16. I firmly believe that every cloud has a silver lining.	1	2	3	4	5
17. My work group is very close.	1	2	3	4	5
18. I distrust the other workers in my department.	1	2	3	4	5
19. I feel like I really belong to my work group.	1	2	3	4	5
20. I don't understand my coworkers.	1	2	3	4	5

[These are sample questions from a comprehensive survey used by Safety Performance Solutions to assess corporate culture.]

Figure 24. Assessing certain person states

> **Scoring for the twenty person-state questions:**
>
> **Self-Esteem** (items 1-4) = feelings of self-worth and value
> (*I'm valuable*). Actual scale = 16 items.
>
>   (a) Add numbers for items 1 & 3.   Total 1 = ____
>
>   (b) Add numbers for items 2 & 4 and subtract from 12.   Total 2 = ____
>
> ---
>
> **Self-Efficacy** (items 5-8 = general levels of beliefs in one's
> competence (*I can do it*). Actual scale = 23 items.
>
>   (a) Add numbers for items 5 & 8.   Total 1 = ____
>
>   (b) Add numbers for items 6 & 7 and subtract from 12.   Total 2 = ____
>
> ---
>
> **Personal Control** (items 9-12) = the extent that a
> person believes he or she if personally responsible
> for his/her life situation (*I'm in control*).
> Actual scale = 25 items.
>
>   (a) Add numbers for items 10 & 11.   Total 1 = ____
>
>   (b) Add numbers for items 9 & 12 and subtract from 12.   Total 2 = ____
>
> ---
>
> **Optimism** (items 13-16) = the extent to which a person
> expects the best will happen for him/her
> (*I expect the best*). Actual scale = 8 items.
>
>   (a) Add numbers for items 15 & 16.   Total 1 = ____
>
>   (b) Add numbers for items 13 & 14 and subtract from 12.   Total 2 = ____
>
> ---
>
> **Belonging** (items 17-20) = the perception of group cohesiveness (*I belong to a team*).
> Actual scale = 20 items.
>
>   (a) Add numbers for items 17 & 19.   Total 1 = ____
>
>   (b) Add numbers for items 18 & 20 and subtract from 12.   Total 2 = ____
>
> ---
>
> **RESPONSIBILITY SCORE** = Sum of Self-Esteem,
> Self-Efficacy, Personal Control, Optimism, and
> Belonging Totals.   **Total Score** = ____
>
> **Figure 24.** Assessing certain person states (*continued*)

You could compare employees' reactions to surveys of perceptions and person states before and after implementing a safety improvement process. Studying reactions prior to an intervention helps identify issues or work areas needing special attention. This information can lead you to choose a particular intervention approach, or customize one. Data from a baseline perception survey might even indicate that a culture is not ready for a certain intervention process, suggesting the need for more education and discussion to get employee "buy-in."

Perception surveys enable accountability for person states.

The impact of an intervention can be measured by comparing perception surveys given before and after implementation. Reviewing the results of such a survey can help people understand the relationship between work practices, perceptions, and attitudes. It can also reveal whether people believe a particular intervention process worked. It's often not enough to be accountable for objective changes in behaviors and environmental conditions. Perception and person state surveys enable us to account for important subjective dimensions relevant to enhancing personal responsibility for safety.

Completing the survey in Figure 24 will help you appreciate the meaning of the person states related to feeling personally responsible for the safety of others. The survey lists 20 items designed to measure five person states. The first four items assess self-esteem, the next four target self-efficacy, the next four — personal control, then optimism, and the last four items measure belonging. Higher scores reflect higher levels of each person state. Research suggests that three of these states — self-efficacy, personal control, and optimism — relate most directly to feeling empowered or responsible.

Please do not consider your results to be valid. Accurate measures of these five person states require completion of much longer surveys. Figure 24 only includes sample items. Also, you need to understand that even an accurate measure of these characteristics would reflect only a state, not a trait. As I've already discussed, states are temporary and can be changed dramatically by environmental conditions and interpersonal relationships.

Accountability for Near Misses

Serious injuries lead to thorough investigations. Environmental conditions and worker behaviors are examined to find causes and take corrective actions. An environmental hazard might be removed, a protective device might be installed, a work procedure might be altered, a training process might be implemented, or a new item might be added to an environmental or behavioral audit process. The result is a reduction in the probability that the injury will occur again.

Do we have to wait until a serious injury occurs before correcting environmental and behavioral conditions diagnostic of a serious injury? Of course not. I'm sure you're well aware of the value of investigating a "near-miss." After all, the only difference between most near-miss experiences and injuries is timing or a few inches. The environmental and behavioral conditions are essentially the same in both a near miss and a serious injury. So, searching for root causes of near-miss experiences and following up with corrective action will lead to lower injury rates.

But there's a problem with near-miss reporting. People don't like to do it. And so opportunities for proactive injury-prevention are missed. Let's explore why it's difficult to turn near-miss experiences into useful safety information. In a nut shell, near misses are hard to pin down. Here's why:

- It's usually inconvenient to fill out a "near-miss investigation form."

- It's convenient and sometimes less stressful to just forget the near miss ever happened.

There are several barriers to near-miss reporting.

- Near-miss experiences are typically private affairs and there's no way to hold people accountable for them. Who wants to report a personal experience that reflects at-risk behavior, inattention and carelessness, and maybe an irresponsible attitude?

- Sometimes the organizational culture deters near-miss reporting. What's to be gained from filing such a report — pleasant or unpleasant consequences? Is the near-miss report considered a positive actively caring attempt to improve safety and health, or an error caused by an uncaring or inattentive worker?

- Slogans like "all injuries are preventable" don't help. Employees think to themselves, "If all injuries are preventable and I almost got injured, I sure don't want anyone to think I'm so careless." Or "If they already know enough to prevent all injuries, they don't need to know about my near miss."

Without a large-scale shift in perception or mindset, near miss experiences will not be readily available for analysis and corrective action. Offering rewards for reporting near-misses would help, but incentives could seem unfair because it's unlikely everyone has an equal chance to file a report. And a person could readily fabricate a near-miss incident to receive a reward. I think the answer to this problem is found in the theme of the latest book by Frank E. Bird, Jr. and George Germain (1997), *The Property Damage Accident: The Neglected Part of Safety.*

Accountability for Property Damage

In their latest book, Bird and Germain stress the importance of environmental assessment, in particular the need to analyze incidents resulting in property damage — but no injuries. In effect, these are near misses. And if damaged equipment or physical structures are not repaired, injuries will eventually follow. Yet the value of investigating property damage incidents is sorely overlooked.

Front-line workers know this for a fact. At workshops and seminars, I hear them talk about stockpiles of broken ladders; tools in disrepair; machine guards that don't work properly; and dents in equipment, walls, and vehicles. This can have dramatic impact on employee's work demeanor, which in turn influences their attitude about safety. For some, property damage signifies that "management doesn't care about our work situation," or "it's okay to damage property as long as we meet production demands."

The Heinrich ratio. Heinrich proposed more than 60 years ago a 300:29:1 ratio between near-miss incidents, minor injuries, and major injuries. Ever since, safety professionals have been encouraged to investigate near-miss incidents in order to reduce minor and major injuries. Heinrich also estimated that 88 percent of all near misses and workplace injuries resulted from unsafe acts.

Heinrich's "law" was an estimate.

It's interesting that the 300-30 ratio of near misses to injuries is referred to as a "law," when in fact it was only an estimate. It wasn't until more than 30 years later that this "law" was actually tested empirically. As director of engineering services for the Insurance Company of America, Frank E. Bird, Jr. analyzed 1,753,498 "accidents" reported by 297 companies. These companies employed a total of 1,750,000 employees who worked more than three billion hours during the exposure period analyzed.

The result was a new ratio: On average, for every 600 near misses, there will be 30 property damage incidents, 10 minor injuries, and one major injury. Now we see the critical link of property damage incidents, which were unidentified in Heinrich's estimates.

Bird found 30 property damage incidents per 10 minor injuries.

Bird and Germain point out that the 600:30:10:1 ratio was obtained from incidents reported and discussed, some during 4,000 hours of confidential interviews by trained supervisors. It's likely the base number is much larger than 600. But notice the dramatic difference between the two ratios. Suppose the number of minor injuries in the

Bird's ratio was multiplied by a factor of three to make it comparable to the 29 minor injuries in the Heinrich ratio. Then you would have 1,800 near misses and 90 property damage incidents per 30 minor injuries. Do you see how linking property damage to workplace injuries can encourage more incident reporting, analysis and corrective action?

Property damage is a physical trace of an incident, and the precursor of an injury. These damages need to be analyzed for their root causes and then repaired. The work culture, including the management system and everyone it encompasses, needs to recognize this and respond appropriately. Behavior that contributed to a property-damage incident should not be considered thoughtless and careless, but failing to report such damage and make repairs shows a lack of personal responsibility for safety.

In Conclusion

Injury prevention requires the identification of environmental conditions and work behaviors that need correction. Regular environmental and behavioral audits provide a mechanism for such an identification and correction process. However, this is not sufficient because it's not possible to define, observe, and correct all possible precursors to an injury. Many environmental conditions or behaviors are not recognized as critical to injury prevention until a certain incident draws attention to at-risk conditions or behaviors.

Near-miss and property damage incidents are injury-diagnostic events. We can learn a lot about injury prevention by investigating the environmental and behavioral factors contributing to these incidents. However, there is severe under-reporting and under-investigation of near misses and damage to property. The near miss implies personal error and is easy to discount. But property damage is physical and countable, and its correction is visible verification that management cares about injury prevention.

Thus, to promote personal responsibility for safety an accountability system needs to do the following: a) be proactive; b) promote the reporting of all injuries, near misses, and property damage incidents; c) distinguish between the journey (process goals) and the destination (purpose or vision); d) hold people accountable for numbers they can control; e) use recognition, rewards, and celebrations to shape process behaviors; f) promote investigation and corrective action for property damage, as well as for near misses and minor injuries; g) consider environment, behavior, and person factors in all incident investigations; and h) include daily audits of work practices and environmental conditions, in addition to periodic assessments of perceptions and person states (through interviews and focus group meetings, as well as plant-wide surveys).

Evaluations of employees' perceptions should be an assessment of whether they feel empowered when it comes to safety. More specifically, the evaluations should address how well the organization's accountability system for safety enables and promotes these person states: 1) self-efficacy ("I can do it"), 2) personal control ("I'm in control"), and 3) optimism ("I expect the best").

What will it take to make you feel more responsible for safety?

Then employees need to be asked, "What will it take to make you feel more empowered or responsible for safety around here?" Seeking frank answers to this question and then struggling to make the changes requested will probably do more than anything else to increase employee involvement, commitment, and responsibility for. The next chapter explains why this is so, and identifies a few key reactions you should expect to get when you ask that question.

Help People Feel Important

Chapter 5

More than sixty years ago, Dale Carnegie (1936) taught us that the key to winning friends and influencing people is to help people feel important. By referring to the scholarship of leading psychologist Sigmund Freud and profound philosopher John Dewey, and by relating one case study after another, Carnegie convinces any reader that the desire to be important is among the most powerful motivators of human performance.

We feel good when we feel important.

You don't have to read Carnegie's classic book to realize the value of this chapter's title. Just reflect on your own life experiences for a moment. Aside from basic physical pleasures, what kinds of situations make you feel really good? I bet many if not most of your good feeling states happen when someone or something made you feel useful, needed, valuable, or significant. These are all synonyms for "important." Many of those times when you don't feel so good psychologically can be traced to a person or event that reduced your sense of importance.

Actually, each of the principles I've already discussed for increasing personal responsibility for safety can be linked to the concept of personal importance. In other words, when we increase peoples' feelings of importance related to their safety performance, we increase the likelihood they will gain some personal responsibility or commitment for safety. So the rationale for decreasing top-down control over safety (Chapter 2) could be based on the reasoning that enforcement or punishment procedures decrease a recipient's sense of importance. And the techniques discussed in Chapter 3 for increasing feelings of empowerment probably also enhance personal perceptions of importance.

It makes sense doesn't it, that whenever you give people knowledge, skills, resources, and authority to do something they believe is important, you will increase their sense of personal value? Psychologists refer to feelings of personal value as "self-esteem." Research I've reviewed elsewhere (Geller, 1996a) has shown a direct relationship between people's levels of self-esteem and their willingness to go beyond the call of duty for the safety or health of another person. In other words, people with higher self-esteem are more apt to actively care.

Increasing self-esteem increases willingness to actively care.

Now consider that our self-esteem or sense of importance varies to some extent moment-by-moment depending upon situational circumstances and interpersonal relationships. What happens to your self-esteem, for example, when someone asks your advice on an important matter, actively listens to your ideas on a critical issue, genuinely acknowledges your accomplishments in a one-on-one encounter, or follows through with your specific improvement suggestions? I'm sure you'll admit that these situations would increase your self-esteem to some degree at least, if only temporarily.

Since we have defense mechanisms to protect our self-esteem, we might not experience a lowered sense of personal value when circumstances indicate lack of personal competence. One of these protective devices is the self-serving bias, as I discussed in Chapter 2. For example, the self-esteem of my university students is not necessarily lowered after they flunk one of my exams. Rather than attribute a failing grade to their own inadequacies, they are more likely to blame factors outside of their control like unfair questions, poor instruction, or friends who kept them up all night.

We protect our self-esteem with the self-serving bias.

So negative consequences or events that reduce feelings of confidence or personal control might not lower our self-esteem and willingness to actively care. But one thing is certain, events or circumstances that decrease our feeling of importance will not increase actively caring behavior. In other words, they won't facilitate the acceptance of personal responsibility for other people's safety and health.

This chapter focuses on two basic approaches to increasing people's self-esteem and thus their willingness to extend their personal responsibility for safety beyond accountability. Several intervention strategies for increasing actively caring for safety fall under these general approaches. One approach is to increase opportunities for individual and group choice, and the other is to increase the significance or meaningfulness of one's safety-related activities. Let's first consider the value of choice.

Increase Opportunities for Choice

We feel important when we feel in control. Offering people choices increases their perception of control and their sense of importance. This in turn enhances their responsibility and motivation to give their best effort. Research has shown that the choice situation does not have to be very complex. In fact, even the simplest choices can make a difference. I found this out more than 25 years ago in a very simple experiment.

A Laboratory Study

In individual one-on-one sessions, I showed half of 40 total subjects a list of five three-letter words (cat, hat, mat, rat, bat) and asked each to select one. Then, after a warning tone, I presented the selected word on a screen in front of the subject, and he or she pressed a telegraph key as fast as possible after seeing the word. The delay in milliseconds between the presentation of the word and the subject's response was a measure of simple reaction time. This sequence of events — warning signal, word presentation, and subject reaction — occurred for 25 trials. If a subject reacted before the stimulus word was presented, the reaction time was not counted, and the trial was repeated.

Choice increases motivation.

The word selected by a particular subject was used as the presentation stimulus for the next subject. Therefore, this subject did not have the opportunity to choose the stimulus word. As a result, the word choices of 20 subjects were assigned (without choice) to 20 other subjects. So, this simple experiment had two conditions — a "choice" condition in which subjects chose a three-letter word for their stimulus and an "assigned" condition in which subjects were assigned the stimulus word selected by the previous subject. To my surprise, the mean reactions of the subjects in the "choice" group were significantly faster than those of the subjects in the "assigned" group.

I interpreted these findings by assuming the opportunity to choose their stimulus word increased motivation to perform in the reaction time experiment. I must admit, however, I didn't expect to find the statistically significant differences. How could the simple choice of a three-letter word motivate faster responding in a simple reaction time experiment? I frankly didn't feel confident in my explanation for these results, and thus I didn't document this research for professional publication. Only later did I appreciate the importance and ramifications of these findings.

From Laboratory to Classroom

About a year after my simple reaction time experiment, I tested the theory of "choice" as a motivator in the college classroom. I was teaching two sections of social psychology: one at 8:00 a.m. Monday, Wednesday, and Friday, and the other at 11:00 a.m. on the same days. There were about 75 students in each class. On the first day of classes, instead of distributing a syllabus with weekly assignments, I distributed only a general outline of the course. This syllabus introduced the textbook, the course objectives, and the basic criteria for assigning grades (a quiz on each textbook chapter and a comprehensive final exam on classroom lectures, discussions, and demonstrations).

I assigned my second class the choices of my first class.

I gave the 8:00 class the following three "choice" opportunities: a) they could choose the order in which the ten textbook chapters would be presented, b) they could submit multiple choice questions for me to consider using for the chapter quizzes, and c) they could hand in short answer and discussion questions for possible use on the final exam. The 11:00 class received the order of textbook chapters selected previously by the 8:00 class in an open discussion and voting process. The 11:00 class was also not given an opportunity to submit quiz or exam questions.

So, I derived "choice" and "assigned" classroom conditions comparable to the two reaction time groups I had studied one year earlier. Posing as regular students, two of my undergraduate research assistants attended each of these classes, and systematically counted amount of class participation. These observers did not know about my intentional choice-assigned manipulations.

From the day the students in my 8:00 class voted on the textbook assignments, they seemed more lively than the students in the 11:00 class. My perception was verified by the participation records of the two classroom observers. Furthermore, the quiz grades, final exam scores, and my teaching evaluation scores from standard forms distributed during the last class period were significantly higher in the "choice" class than the "assigned" class. Although several students from the 8:00 class submitted potential quiz and final exam questions, I didn't use any of these questions. Each class received the same quizzes and final exam. The final grades were significantly higher in the 8:00 class than the 11:00 class.

Choice increased involvement and learning.

There are several possible reasons for the group differences. I'm convinced the "choice" versus "assigned" manipulation was a critical factor. I believe the initial opportunity to choose reading assignments increased class participation and this involvement fed on itself and led to more involvement, choice, and learning. The students' attitudes toward the class improved as a result of feeling more "in control" of the situation rather than "controlled." It's likely the "choice" opportunities in the 8:00 class were especially powerful because they were much more different than the traditional top-down classroom atmosphere, as typified by my 11:00 class. In other words, the contrast with their other courses made the choice opportunities especially salient and powerful.

Practical Implications

The implications of these "choice" versus "assigned" findings in laboratory and classroom investigations are far reaching. The notion that "choice increases involvement" relates to a number of motivational theories supported by psychological research. Essentially, when people believe they have personal control over a situation they are generally more motivated to achieve, and they get more involved (Steiner, 1970; White, 1959). It has also been found that the debilitating effects of environmental stressors are lessened when people feel a sense of personal control in the situation. In fact, stress can be positive when we stay "in control." In this case, the stressor merely motivates us to work harder to stay "in control."

The difference between stress and distress is personal control.

When we lose a sense of personal control following an environmental stressor, we experience "distress." This state is debilitating, and can lead to burnout or learned helplessness. The key point here is that personal control increases personal responsibility and decreases the possibility of stress turning to distress or burnout. And, how can this sense of personal control be increased? You guessed it. We can increase a belief in personal control by increasing the number of "choice" opportunities in a situation.

Opportunities to choose lead to increased personal responsibility and more involvement, and more involvement leads to increased perceptions of personal responsibility. And more personal responsibility leads to more choice and more involvement — and this continuous involvement cycle continues. This is a primary route to true empowerment as discussed in Chapter 4.

Past experiences influence acceptance of current choice opportunities.

Some people's past experiences make them less likely to develop a sense of personal control when "choice" is offered. They may mistrust the "choice" situation, or lack confidence in their ability to make something positive come out of the opportunity. They might not feel comfortable with the added responsibility of "choice," and thus resist the change implied by new "choice" opportunities.

Usually the best way to deal with this resistance is not to confront it directly. Hopefully, enough other people will take personal control of their choice opportunities and eventually convince the resisters to get involved. It is important, however, that people with "choice" feel competent to make appropriate decisions, and therefore education and training might be needed. The bottom line is that people want to be seen as responsible and in control of important circumstances. It just may take a little time to convince some people to believe a certain safety-related situation is important and worthy of their "choice" behavior.

Choice is motivating when it makes a difference.

The consequences of choice are critically important. If the subjects in my simple reaction time experiment or my social psychology class did not believe their choices made a difference in the situation, the choice opportunities would not have made a difference in their personal responsibility and motivation. The reaction-time subjects saw me use their choice in the stimulus presentations, and the psychology students observed me change the class structure and process as a result of their choices. When we see a consequence consistent with decisions from our choice opportunities, we increase our trust of the people who gave us the power to choose and we gain confidence in our abilities to take personal control of the situation.

Choice for Corporate Safety

An Exxon Chemical facility with 350 employees used the power of choice to eventually get all employees involved in a behavior-based safety process. The employees initiated an actively caring observation, feedback, and coaching process in 1992, reaping amazing safety benefits for their efforts. In 1994, for example, 98% of the workforce had participated in behavioral observation and feedback sessions, documenting a total of 3,350 coaching sessions for the year. A total of 51,408 behaviors were systematically documented on critical behavior checklists, of which 46,659 were safe and 4,389 were at risk.

The comprehensive employee involvement in a behavioral observation and feedback process at this Exxon Chemical facility led to remarkable outcomes. At the start of their process in 1992, the plant safety record was quite good (i.e., 13 OSHA recordables for a TRIR of 4.11). The record was improved to five OSHA recordables in 1993 (TRIR=1.60), and in 1994 the facility had the best safety performance in Exxon Chemical with only one OSHA recordable (TRIR=0.35). In 1997 this plant reached an enviable milestone — they had an injury-free workplace for an entire year. In 1998, they were still going strong.

Choosing the procedures for behavior-based safety increases ownership.

I've seen many companies improve their safety performance substantially with processes based on the principles of behavior-based safety, but this plant holds the record for getting *everyone* involved and for obtaining exceptional results. I'm convinced a key factor in this organization's outstanding success was the employees' "choice" in the development, implementation, and maintenance of the process. From the start, the employees owned their behavioral observation and feedback process because they applied behavior-based principles *their way*. Here's what I mean.

There is no "best" way to implement behavior-based safety; rather the principles and procedures of behavior-based safety need to be customized to fit the target work culture. And, the most efficient way to make this happen is to involve the target population in the customization process. At this Exxon Chemical facility, the entire workforce learned the behavior-based principles by participating in ten, one-hour small-group sessions spaced over a six-month period.

Employee choice is fundamental to a most successful behavior-based safety process.

The education/training sessions were facilitated by other employees who had received more intensive training in behavior-based safety. At these group sessions, employees discussed specific strategies for implementing a plant-wide behavioral monitoring and coaching process, and they entertained ways to overcome barriers to total participation and sustain the process over the long term. They designed a process which includes employee choice at its very core. Although many specifics of the process have changed since its inception in 1992, the choice aspect remains a constant.

Each month, employees schedule a behavioral observation and feedback session with two other employees — safety observers. They select the task, day and time for the coaching session, as well as two individuals to observe their performance and give them immediate and specific feedback regarding incidence of safe and at-risk behaviors. Employees choose their observers (and coaches) from anyone in the plant. At the start of their process the number of volunteer safety coaches was limited to about 30% of the workforce, but today everyone is a potential safety coach. Personal choice facilitated responsibility, involvement, ownership, and trust in the process.

At first, some employees did not have complete trust in the process and resisted active participation. Some tried to "beat the system" by scheduling their observation and feedback sessions at slow times when the chance of an at-risk behavior was minimal, such as when they were watching a monitor or completing paperwork. But most employees were certainly "on their toes" when the observers arrived at the scheduled times. At the same time, those observed were optimally receptive to constructive feedback and advice from the observers they had selected. Many people (whether observing or being observed) were surprised to find numerous at-risk behaviors occurring in situations where employees understood the safe operating procedures and knew they were being observed for the occurrence of at-risk behaviors.

Choosing to observe (as a coach) and to be observed (as a recipient of feedback) builds personal responsibility.

Today, most employees at this facility schedule their coaching sessions during active times when the probability of an at-risk behavior is highest. Frequently, the observed individual uses this opportunity to point out an at-risk behavior necessitated by the particular work environment or procedure, such as a difficult-to-reach valve, a hose-checking procedure too cumbersome for one auditor, a walking surface made slippery by an equipment leak, or a difficult-to-adjust machine guard.

Thus, many employees have chosen to use their observation and feedback process to demonstrate that some at-risk behaviors are actually facilitated or required by equipment design or maintenance, and by environmental conditions or operating procedures. This often leads to workers taking on the personal responsibility of completing a safety work order or initiating specific improvement in environmental conditions or operation procedures.

From Choice to Importance

I hope you're convinced that it's useful to increase people's sense of importance by giving them opportunities to gain personal responsibility through choice. Actually, you probably didn't need to hear about my personal research and a case study to believe in the power of choice. You need only reflect on your own life circumstances to realize how a perception of choice or personal control increases motivation, involvement, and personal responsibility.

We're not always in control of the critical events of ongoing life events, and thus we've experienced the frustration, discomfort, and distress of being at the mercy of environmental circumstances or another person's decision. You've certainly experienced the pleasure of having alternatives to chose from and feeling in control of those factors critical for success. How sweet is the taste of success when we can attribute the achievement to our own doing!

Increase importance through choice opportunities and you'll increase personal responsibility.

I hope the message is clear. Increase people's importance by giving them the power to choose safety procedures consistent with the right principles, and the result will be more responsibility, involvement, and ownership. This may require relinquishing some top-down control, abandoning a desire for a "quick fix," changing from focusing on outcomes to recognizing process achievements, and giving people opportunities to choose, evaluate, and refine their means to achieve the ends. It might also require convincing people their choice of certain alternatives is really important. In other words, the value of having choice depends upon people believing in the importance of the task or job for which they are choosing. This brings us to the next topic — increasing the perceived importance of safety-related performance.

Assure the Significance of Safety-Related Behavior

We increase our sense of importance when we believe we're doing something important. I'm convinced many if not most corporate cultures, especially the management system, place more importance on productivity and quality than on safety. Thus, it's no wonder safety often takes a back seat to production. Not only does production have an achievement-focused scoring system compared to the loss-control perspective of safety (as discussed in Chapter 3), production is typically viewed as being more important than safety. After all, if the product doesn't get out the door, the company doesn't make a profit and employees don't get paid.

Production is commonly viewed as being more important than safety.

The financial consequences of a business are directly connected to quality and quantity of product output. We hear about the costs of property damage and personal injury. But the connection between these losses and corporate profit is not easy to see on a daily basis. When you're on the job, you see how efficient work behavior leads to productivity and then to financial gain. You can even see how taking a short cut or calculated risk enables faster output and more potential earnings. This can give you a feeling of importance. You made something happen faster or more efficiently, and thus optimized the production system. Your special skills put you in control of the process and the company's profits. Your job is important and so are you.

When it comes to safety, however, your attitude might be different. Not only is it uncomfortable, inconvenient, or inefficient to follow the various safety rules, the extra procedures detract from the bottom line — quality output delivered on time. So the safe way is not the most productive way, and therefore does not rank high in importance.

Obviously, this perspective needs to be turned around if we want people to feel more responsible for safety. How can we do that? How can we increase the perception that safety-related behaviors are important, and therefore we are important when we perform them? Better yet, how can we convince people that teaching or encouraging others to make safe decisions is *extremely* important? If promoting safety is viewed as important, then people will increase their sense of importance whenever they help others actively care for safety or health. This builds personal responsibility for safety while also contributing to achieving an injury-free workplace.

How can we convince people that actively caring for safety is important?

Of course we can easily convince ourselves that preventing personal injury is much more important than making money. Is anything more important than leaving work in the same physical state as when you arrived? Most people agree that being in good physical health is the most important blessing to be thankful for. The trouble is, most of us take our health and safety for granted. Plus, it's not easy to see how refraining from taking calculated risks protects our health and safety. We usually get away with taking calculated risks to be more efficient or productive. In this case, our at-risk behavior is viewed as important, efficient, and "safe."

Safety responsibility requires impulse control.

So being accountable and responsible for safety requires impulse control under the most difficult circumstances. Safety leaders hold people accountable for doing certain things that are uncomfortable, inconvenient, or inefficient in order to avoid a negative consequence that seems remote and improbable. Asking people to do this so they feel responsible takes a special kind of intelligence, more critical for lifetime success than any other form of intelligence. And avoiding efficiency for safety takes the same kind of intelligence.

Consider that promoting safe productivity over efficient output builds this special kind of intelligence among both the agents and recipients of a safety intervention. In other words, when we ask people to avoid a calculated risk in order to achieve a delayed and remote positive consequence (avoiding an injury), we increase this intelligence in ourselves. And when these people willingly follow your safety suggestions they are enhancing this sort of intelligence in themselves. This special type of human intelligence is strengthened whenever we accept more personal responsibility for safety.

Actively caring for safety builds a most important kind of intelligence.

The bottom line here is that people who go beyond the call of duty for safety should feel especially important because they enhance a special kind of intelligence in themselves and in others. So taking personal responsibility for safety does more than prevent injuries. It contributes to cultivating a unique form of intelligence throughout a culture. Realizing this can be quite motivating, so let's turn to a discussion of this kind of intelligence and its relationship to industrial safety. Learning about this exceptional brand of intelligence enabled me to increase my own sense of personal importance and responsibility.

A Personal Story

It happened over 40 years ago, but I remember the incident as if it were yesterday. The 27 students in my sixth-grade class took an intelligence test, and then we followed the teacher's instructions to grade our own exams. The teacher announced that a score of 100 meant average intelligence. She said that those of us who didn't score higher than 100 should consider a career path that did not require a college education. I was devastated — my score was exactly 100. From the time I entered the first grade I had my heart set on becoming a medical doctor like my father, but now my test score said I was not smart enough even to attend college. When I asked the teacher later, she said I should lower my aspirations and consider vocational education.

Fortunately I had supportive parents who helped me get over this painful event, and insisted I should follow my dreams. They reassured me I should not take the results of that test seriously and pointed out my other achievements. After all, I was an "A" student and was successful in a variety of other group and individual activities, from sports and Boy Scouts to playing a musical instrument. Still, it wasn't until I studied psychological testing in college that I realized the fallacy of that traumatic sixth-grade testing experience.

Obviously, students should not have been allowed to see their test scores. In fact the self-scoring of aptitude and achievement tests ceased many years ago. My education also informed me about the relative effectiveness, or I should say ineffectiveness, of intelligence tests to predict success in college and a future career. Most importantly, my later study of research in graduate school revealed another kind of intelligence which is much more predictive of people's ability to follow their dreams. And the exciting thing about this intelligence is that it can be learned and developed through education, training, and interpersonal communication. Daniel Goleman (1995) calls this special ability "emotional intelligence" in his best selling book by the same title.

What is Emotional Intelligence?

First, it's important to realize the fallacy of the intelligence quotient (IQ) as a key indicator of future success. When I was an adolescent in the 1950's, a person's IQ or mental capacity was considered a key determining factor of success in school and in a later career. It was believed you had to have a high IQ in order to be a successful engineer, doctor, lawyer, or university professor. And, it was commonly believed in those days that paper and pencil tests were available to obtain a fair and accurate measure of a person's intelligence. For example, your score on the Scholastic Aptitude Test (SAT) was presumed to measure your capacity to handle college-level course work.

Emotional IQ is key to personal success.

Today, it is generally believed that intellectual capacity (or IQ) is much more complex and difficult to measure. And the numbers obtained from intelligence tests (including the SAT) are not very effective at predicting success in college or in a professional career. Not too long ago the "A" in SAT was referred to as "Achievement" rather than "Aptitude" to reflect a measure of one's level of knowledge from personal learning experiences rather than an inborn intellectual capacity. Today, the Educational Testing Service (creators of the SAT) use "Assessment" as the A-word, presumably to reflect a middle ground between aptitude (innate ability) and achievement (acquired knowledge). The important point for our discussion is that the SAT (whatever the "A" word) is often a terrible predictor of success (or failure) in college, and this is largely because of "emotional intelligence."

Figure 25. Optimism means we expect the best.

Chapter 5 — Help People Feel Important

Figure 26. Interpersonal intelligence affects intrapersonal intelligence, and vice versa.

From his comprehensive review of the research, Daniel Goleman concludes that "At best, IQ contributes about 20 percent to factors that determine life success, which leaves 80 percent to other factors" (p.34). He then goes on to show convincing evidence that a majority of the other factors can be associated with "emotional intelligence" or one's ability to: a) remain in control and optimistic following personal failure and frustration, and b) to understand and empathize with other people and work with them cooperatively. Howard Gardner (1993) refers to the first ability as "*intra*personal intelligence" and the second as "*inter*personal intelligence."

We can increase emotional IQ in ourselves and others.

Intrapersonal IQ. We show intrapersonal intelligence when we keep our negative emotions (including frustration, anger, sadness, fear, disgust, and shame) in check and use our positive emotions or moods (such as joy, passion, love, optimism, and surprise) to motivate constructive action. Much of my motivation to study for exams or prepare for speeches, for example, comes from a strong desire to avoid the negative emotions (such as fear and frustration) associated with being unprepared to handle a stressor. And when such preparation leads to success, the joy of achievement builds self-confidence, personal control, optimism, and personal responsibility. Then, these pleasant consequences motivate hard work to achieve them again. Figure 25 illustrates how optimism influences people to expect the best. And when we expect the best we are more likely to do what it takes to make the best of existing circumstances.

Interpersonal IQ. We demonstrate *inter*personal intelligence when we correctly recognize the moods, emotions, motives, or feeling states of other people and react appropriately. As I cover in my recent book on teamwork (Geller, 1998), this kind of emotional intelligence requires both active listening and behavior-based feedback. Thus, people with high interpersonal intelligence communicate with other people to increase their self-confidence, personal control, optimism, and personal responsibility. When we communicate with interpersonal intelligence, we facilitate the cultivation of intrapersonal intelligence in others. As illustrated in Figure 26, interpersonal communication can reduce or enhance intrapersonal intelligence. And the process of interpersonal communication is reciprocal and mutually supportive of constructive or destructive emotional states.

Safety, Emotions, and Impulse Control

Safety professionals need to develop emotional intelligence in themselves and others. Think about the range of emotions that come into play as we struggle to improve workplace safety and health. We need the curiosity to assess objectively the impact of our safety interventions, persistence to continue successful programs in the face of active resistance, flexibility to try new approaches, resilience to bounce back after failure, and passion to try again.

Achieving the vision of an injury-free workplace requires awareness and control of our own emotions, as well as the ability to assess, understand, and draw on the influence of other people's emotions. This requires empathic and persuasive communication skills (interpersonal intelligence), as well as self-confidence, personal control, self-esteem, and optimism (intrapersonal intelligence) to develop and implement new tools for safety management.

Taking personal responsibility for safety builds emotional intelligence, and vice versa.

Perhaps we can increase personal responsibility for safety by helping people understand the fundamental emotional problem at the root of all safety intervention. Safety requires impulse control under the most difficult circumstances. We ask employees to do things that are uncomfortable or inconvenient in order to avoid a negative consequence that seems remote and improbable. This takes a special kind of emotional intelligence, both from us as safety professionals and from the employees we're working with.

Dr. Goleman considers impulse control "the root of all emotional self-control" (p.81), and he demonstrates its power in the classic research of Walter Mischel, a renowned psychology professor at Stanford University. In the 1960s, Dr. Mischel gave four-year-olds a "marshmallow test" to measure impulse control. In the test, children were given a marshmallow and told they could eat it immediately or wait until later and receive two marshmallows. Some children ate the single marshmallow within a few seconds after the researcher left the room; others were able to wait the 15 to 20 minutes for the researcher to return.

The "Marshmallow Test" measured impulse control.

The diagnostic power of this simple test was shown when these preschoolers were followed up as adolescents. Those who put off immediate gratification for a bigger but delayed reward demonstrated greater intrapersonal and interpersonal intelligence. They handled stressors and frustration with more confidence, personal control, and optimism. They were more self-reliant, trustworthy and dependable, and less likely to shy away from social contacts than the children who had not waited for two marshmallows at age four.

In comparison, the adolescents who had devoured the single marshmallow 12 to 14 years earlier were now more stubborn and indecisive, more prone to jealousy and envy, and more readily upset by stress or frustration than the adolescents who had waited for the extra marshmallow.

The "Marshmallow Test" predicted future success.

When evaluated again during their last year of high school, those who had waited patiently at age four were far superior as students than those who failed the marshmallow test. They were clearly more academically competent; they had better study habits, and appeared more eager to learn. They were better able to concentrate, to express their ideas, and to set goals and achieve them. Most astonishingly, these higher achievers scored significantly higher on both the math and verbal portions of the SAT (by an average of 210 total points) than the students who had not delayed gratification at age four.

Nurturing Emotional Intelligence

Although Mischel's research and Goleman's conclusion suggest that some degree of emotional intelligence begins early in life, there is plenty of evidence that emotional intelligence (both intra- and interpersonal) can be learned. Goleman describes a number of educational/training programs that have demonstrated success at increasing the emotional intelligence of children. In my book *The Psychology of Safety* (Geller, 1996a), I review a variety of techniques for improving certain feeling states among adults (including self-confidence, personal control, optimism, belonging, and self-esteem) which imply an increase in intrapersonal or interpersonal intelligence.

Safety professionals need intrapersonal and interpersonal intelligence.

I'm sure you see the relevance of emotional intelligence to improving occupational health and safety. Obviously, safety professionals need to remain self-confident and optimistic (intrapersonal intelligence) in their attempts to prevent injuries, and much of their success depends upon their ability to facilitate involvement, empowerment, and win/win cooperation among those who can be injured (interpersonal intelligence). But it's easy for safety professionals to get discouraged and frustrated, because so often safety seems to take a back seat to seemingly more immediate demands like meeting production quotas and quality standards. Controlling these negative emotions is reminiscent of Walter Mischel's "Marshmallow Test."

Doing things for safety (from using protective equipment to completing behavioral and environmental audits) is equivalent to asking someone to delay immediate gratification for the possibility of receiving a larger reward (preventing a serious injury). In other words, safety often (if not always) requires people to control their impulse to procure an immediate consequence (if only to be more comfortable or to complete a task faster). Note, however, that in the "Marshmallow Test" the delayed and larger consequence of two marshmallows was certain and positive. If the researcher has one marshmallow now, he can certainly deliver two later. But with safety the remote consequence of an injury is not only uncertain, it is negative.

In safety promotion the delayed consequence is uncertain and negative.

While promoting safety is analogous to the "Marshmallow Test" of impulse control, it is far more challenging to convince people to delay certain and immediate gratification in order to avoid an uncertain negative consequence than to earn a delayed but certain positive consequence. Plus, our attitude toward a task (or emotional state) is more positive when we are working to gain a pleasant consequence than when working to avoid a negative consequence. This is yet another reason why safety promotion is so challenging, and why it is so important for safety professionals to nurture intra- and interpersonal intelligence in themselves and others. The special significance of meeting this challenge is obvious.

In Conclusion

This chapter focused on two general approaches to increasing people's sense of personal responsibility: increasing opportunities for choice and helping people feel important. The interdependence of these methods is obvious. When people have choices in a situation they feel more important, and when their sense of importance is increased they want to get more involved and make more choices. So each of these approaches to increasing personal responsibility is mutually supportive and motivational.

Choice opportunities increase a sense of importance and personal responsibility.

A behavior-based safety process allows for many opportunities to involve teams of co-workers in a choice or consensus-building process, from selecting critical behaviors to target to developing intervention procedures for increasing safe behaviors and decreasing at-risk behaviors. In this regard, it's necessary to teach all company employees or associates the basic principles of behavior-based safety (as reviewed in Chapter 1). But it is also necessary to encourage teams to customize specific procedures for their work areas. This choice process will be motivating in itself, and will help develop a sense of ownership for the methods and tools. This leads naturally to both personal and interpersonal responsibility to make the process work.

This choice concept must also be considered throughout peoples' daily implementation of a behavior-based safety process, whether the focus is on behavior-based incentives, group observation and feedback, incident analysis, or one-on-one coaching. That is, individual participation should be voluntary. An expectation or desire for total involvement can be expressed, and should be the ultimate aim. But it's essential that people feel their participation is voluntary. If they get involved because they believe they must, their responsibility will likely not exceed this accountability. If they choose to participate, they will likely do more than an accountability system mandates.

Increased emotional IQ is a soon, certain, and positive consequence of behavior-based safety.

People feel more important when given opportunities to make decisions about an important assignment. So it's crucial that people understand the value of behavior-based safety in preventing injuries. That's not easy because the linkages between behavior change, property damage, and personal injury are not obvious. People need to remind themselves and others that while their extra effort for safety might not result in soon, certain, and positive consequences, they are working for a bigger delayed benefit.

The ability to put off soon and certain rewards for ones that are remote and uncertain is called emotional intelligence. And perhaps the greatest value of going beyond the call of duty for safety is that such efforts build emotional intelligence among people. That's a soon, certain, and extremely significant consequence of behavior-based safety that should increase relevant personal responsibility in all of us.

Cultivate Belonging and Interpersonal Trust

Chapter 6

In his best seller, *The Different Drum: Community Making and Peace*, M. Scott Peck (1987) challenges us to experience a sense of true community with others. We need to develop feelings of belonging with one another regardless of political preferences, cultural backgrounds, and religious doctrine. We need to transcend our differences, overcome our defenses and prejudices, and develop a deep respect for diversity. Dr. Peck claims we must develop a sense of community or interconnectedness with one another if we are to accomplish our best and ensure our survival as human beings.

A sense of belonging leads to interpersonal trust and caring.

It seems intuitive that building a sense of community or belonging among our coworkers will increase people's responsibility to organizational safety. Safety improvement requires interpersonal observation and feedback, and for this to happen people need to adopt a collective win/win perspective instead of the individualistic win/lose orientation common in many work settings. A sense of belonging and interdependency leads to interpersonal trust and caring — essential features of a Total Safety Culture.

In my numerous group discussions with employees on the belonging concept, someone inevitably raises the point that a sense of belonging or community at their plant has decreased over recent years. "We used to be more like family around here" is a common theme. For many companies, growth spurts, continuous turnover — particularly among managers — or "lean and mean" cutbacks have left many employees feeling less connected and trusting. Some employees have seemingly regressed from satisfying belonging needs to concentrating on maintaining job security, in order to keep food on the table.

Has belonging and interpersonal trust decreased in your organization?

Figure 27 lists a number of special attributes prevalent in most families, where belonging and interpersonal trust are often optimal. We feel personally responsible for the members of our immediate family. As a result we are very willing to go beyond the call of duty and actively care for the safety and health of our family members. To the extent we follow the guidelines in Figure 27 among members of our "corporate family," we will build responsibility for safety and achieve a Total Safety Culture. Following the principles in Figure 27 will cultivate belonging and trust among people, and lead to the quantity and quality of actively caring behavior expected among family members — at home and at work.

> - **We use more rewards than penalties with *family* members.**
> - **We don't pick on the mistakes of *family* members.**
> - **We don't rank one *family* member against another.**
> - **We brag on the accomplishments of *family* members.**
> - **We respect the property and personal space of *family* members.**
> - **We pick up after other *family* members.**
> - **We correct the at-risk behavior of *family* members.**
> - **We accept the corrective feedback of *family* members.**
> - **We are our brothers/sisters keepers of *family* members.**
> - **We actively care because they're *family*.**
>
> **Figure 27.** Incorporating an actively caring *family* perspective in an organization will lead to a Total Safety Culture.

Increasing Belonging

When people work together on a project they have a sense of togetherness and interdependency. They depend on each other to complete certain components of a task, and they celebrate as a group when goals are reached. In other words, teamwork builds belonging and interpersonal responsibility. Therefore, interventions which help to build successful safety teams also increase a sense of belonging and interpersonal trust. My book, *Building Successful Safety Teams: Together Everyone Achieves More* (1998), presents step-by-step methods and tools for developing effective teams, from selecting team members and translating an assignment into goals and task assignments to maintaining team progress and evaluating results. I refer you to that product, which includes an instructional videotape, facilitator guide, and employee handbook, for a comprehensive discussion of building belonging through teamwork.

Actively Caring States

Figure 28 depicts an actively caring model I have described elsewhere (Geller, 1996a, 1997c) to summarize conditions and situations that influence people's willingness to take personal responsibility for safety. Supported by substantial research, the model predicts that people's willingness to accept responsibility for another person's health or safety increases directly with the enhancement of three basic person states: empowerment, self-esteem, and belonging.

Empowerment is covered in Chapter 4, including ways to enhance feelings of empowerment and thus personal responsibility for safety. Self-esteem is covered in Chapter 5, with a focus on allowing people to feel important which in turn enhances their self-esteem and their willingness to go beyond the call of duty for another person's safety or health. In this chapter we consider the role of belonging and interpersonal trust.

At some of my workshops and seminars, participants express concern that the actively caring person-state model may not be practical. Typical comments are, "Those are just warm fuzzy ideas with no real-world applications," "This sounds like mumble jumble psychology to me," and "That sure sounds good but what can I do about it?"

Participants readily accept the methods and tools of behavior-based safety because they are straightforward, objective, and obviously applicable to the workplace. But these person-based concepts seem subjective, "touchy-feely," and difficult to work with, espe-

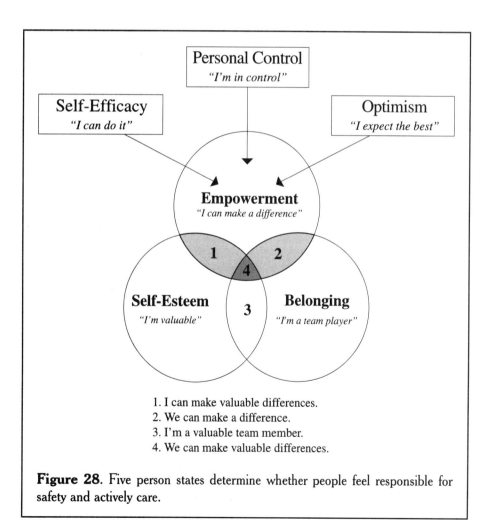

1. I can make valuable differences.
2. We can make a difference.
3. I'm a valuable team member.
4. We can make valuable differences.

Figure 28. Five person states determine whether people feel responsible for safety and actively care.

cially on an organizational level. "How can we get our arms around these warm fuzzies and use them to increase personal responsibility for safety?"

People can identify situations and events that influence their self-esteem, empowerment, and belonging.

These person states are more difficult to define, measure, and manage than behavior. However, they do exist inside all of us, and they do influence what we do. And our behavior influence these feeling states. Furthermore, people can identify relationships between these person states, and environmental conditions and interpersonal behaviors. They can define quite clearly what situations, events, and behavioral consequences enhance versus depreciate their self-esteem, empowerment, and belonging. In fact, that's how the practical guidelines given in Chapters 4 and 5 were derived for enhancing empowerment and self-esteem, respectively.

After introducing the person states depicted in Figure 28 at my two-day workshops, I often ask participants to discuss the person states in groups of five to six, and suggest some practical ways to influence a particular feeling state. This exercise not only promotes profound understanding of each person state, but also convinces people they can get something useful from these "warm fuzzies." In fact, every group I've worked with turned this assignment into realistic recommendations for boosting a particular person state in their work setting.

The most common proposals given by my seminar discussion groups for creating and sustaining an atmosphere of belonging among employees include:

- Use self-managed or self-directed work teams (as explained in Geller, 1998).
- Decrease top-down directives and "quick-fix" programs.
- Increase team-building exercises, group goal-setting, and group feedback.
- Track team progress on visible charts or graphs.
- Teach employees strategies for setting team goals, running effective team meetings, promoting team progress, and evaluating team effectiveness (all covered in Geller, 1998).
- Provide time and resources for team meetings and team exercises.
- Increase group celebrations for both process and outcome achievements (as described in Geller, 1997b,c).
- Develop and implement strategies for increasing interpersonal trust (as detailed in Geller, 1998 and reviewed below).

Understanding Interpersonal Trust

"No way," workers sometimes tell me when I introduce a behavior-based observation and feedback process. They don't like the idea of a coworker watching them and tallying safe and at-risk behaviors for a feedback session. Managers and safety directors will sometimes agree with the basics of the behavior-based coaching process, but then say, "Our plant is not ready for such a process."

Why the resistance? It all comes down to one word — trust. A behavior-based observation and feedback process won't succeed without a high degree of trust. First you must believe the process is valid, a good tool for improving safety. Then you need to have confidence that your coworkers will know how to observe and give feedback correctly. Beyond ability, you must trust that the person observing and recording your behaviors doesn't have a hidden agenda. This is called trusting the intentions of an observer, and it's critical. Also critical is trusting the intentions of your company's managers who are responsible for bringing the process into your workplace.

Behavior-based observation and feedback require a context of interpersonal trust.

How do you know if your organization is ready to trust the behavior-based process, if coworkers will trust each other as coaches, and if enough belonging exists between management and the workforce? After reading this chapter I hope you'll have a better understanding of the feelings that go into trust.

What is trust? "Confidence in the integrity, ability, character, and truth of a person or thing" is the first definition of "trust" given in my *American Heritage Dictionary* (1991). This definition refers to behavior (as in "ability") as well as internal or person-based dimensions (as in "integrity" and "character"). These are the characteristics my colleagues and I measure to assess trust between people in a work culture, using a survey based on research by Cook and Wall (1980).

Interpersonal trust refers to intentions or abilities.

The survey distinguishes between intention and ability. In other words, you can be confident a person means well, but you might doubt his or her ability to complete the intended task. In this case, you trust the individual's intentions but you're not so sure things will work out as planned. You lack confidence in the person's ability to make good on his or her promise.

This is a common perception of workplace safety efforts. Well-intentioned managers or safety leaders can talk about missions, goals, or policies that employees view as idealistic or unrealistic. And when punishment or reward is not carried out consistently and fairly, employees don't trust the abilities of those running the program. The reverse also happens. Employees might have faith in the ability of others, but mistrust their intentions. When a policy change is sprung on workers without warning or rationale, they might suspect management's intentions. They might believe managers have the intellect and skills to make things happen, but will they do the right thing? "Do they really have our welfare in mind?" suspicious minds ask.

What if management sets up a safety incentive program that offers everyone a prize if no one gets hurt for six months? Employees feel pressure not to report an injury. If they do, everyone loses their reward. So what is management's true agenda? Is it to keep the numbers down, or to keep employees from getting hurt?

An outcome-based incentive/reward program can decrease trust in management's intentions.

The same issues of trust play out every day between coworkers. You might have confidence in a coworker's ability to perform a job safely and competently, but you might not want to confide in that person. "My partner might use the information against me to get the promotion before me," could be your thought. Similarly, you might trust the intentions of a coworker ("He would never take advantage of me") but lack confidence in his ability ("I'm sure he'll try his best, but he just doesn't have enough experience").

As you can see, there are four dimensions to trust. These four dimensions were used by Cook and Wall (1980) to develop a questionnaire asking for responses to 12 statements in order to assess "interpersonal trust at work." The instructions for this survey tool and the actual items are given in Figure 29.

The statements below express opinions that people might hold about the confidence and trust that can be placed in others at work, both fellow workers and management. Circle the numbers next to each statement to indicate how much you agree with it.

	Highly Disagree	Disagree	Not Sure	Agree	Highly Agree

Faith in Intention of Peers

1. If I got into difficulties at work I know my co-workers would try and help me out.   1 2 3 4 5

2. I can trust the people I work with to lend me a hand if I need it.   1 2 3 4 5

3. Most of my coworkers can be relied upon to do as they say they will.   1 2 3 4 5

Faith in Intention of Management

4. Management in my company is sincere in its attempts to meet the workers' point of view.   1 2 3 4 5

5. I feel quite confident that the company will always treat me fairly.   1 2 3 4 5

6. Our management would be quite prepared to gain advantage by deceiving the workers.   1 2 3 4 5

Confidence in the Ability of Peers

7. I have full confidence in the skills of my co-workers.   1 2 3 4 5

8. Most of my fellow workers would get on with their work even if supervisors were not around.   1 2 3 4 5

9. I can rely on other workers not to make my job more difficult by careless work.   1 2 3 4 5

Confidence in the Ability of Management

10. Our company has a poor future unless it can attract better managers.   1 2 3 4 5

11. Management can be trusted to make sensible decisions for the company's future.   1 2 3 4 5

12. Management at work seems to do an effective job.   1 2 3 4 5

Figure 29. A survey to assess interpersonal trust.

The researchers originally gave this survey to employees by reading them each item and asking them to indicate on a seven-point scale their degree of agreement with a particular statement. My colleagues and I include these 12 statements in a questionnaire along with many others to draw conclusions about a work culture. We ask respondents to give their opinions on a five-point scale, as shown in the figure.

Except for statements #6 and #10, the higher scale value equates with greater interpersonal trust. Statements #6 and #10 are negatively phrased, and need to be reverse scored. By totaling the scores for all 12 statements (with #6 and #10 reverse scored) you have an estimate of an individual's perception of interpersonal trust in his or her work culture. You can obtain an overall trust index by calculating the mean survey score from many respondents.

There is nothing special or magical about the wording of these statements. Feel free to reword statements if different language fits better with your culture. And you can add or substitute new statements. One more point: This scale was developed to measure interpersonal trust from the viewpoint of an hourly worker. With only slight adjustments, the scale can also estimate interpersonal trust from a manager's perspective.

Once you arrive at an estimate of the level of trust in your workplace, the question becomes: How can we increase it? After all, safety improvement depends on a sense of belonging and interdependence among coworkers and managers. Everyone should look out for the safety of everyone else. This means trusting each others' intentions and abilities to be responsible and do the right thing. Let's consider ways to increase interpersonal trust.

How to Build a Trusting Culture

How can we increase levels of trust at work to improve safety? This is a critical question, specifically when it comes to building personal responsibility for safety and getting the most out of a behavior-based observation and feedback process. For workers to accept feedback, they need to trust an observer's capability (accuracy) and intentions (caring, not punishing). Finding little real-world research on building interpersonal trust, I called together a group of my research students and colleagues and asked them to brainstorm and reach a consensus on how to build trust. After almost two hours, we arrived at an interesting and seemingly useful list of proposals. We also learned something first-hand about developing feelings of trust. This brainstorming process brought us together as a group that relied on each other for ideas and feedback.

I urge you to conduct a consensus-building exercise similar to the one we used. Your ideas will likely be similar to ours but the recommendations will be owned by your group. And people are more likely to follow recommendations they had a hand in developing.

We reached a consensus that the following words — all beginning with the letter "C" — capture the essence of trust building. Use them to kick off your own discussion of trust.

Body language speaks more than words.

Communication. What people say and how they say it influences our trust in both their capability and their intentions. I'm sure you've heard many times that how something is said, including intonation, pace, facial expressions, hand gestures, and overall posture, has greater impact than what is actually said.

There is probably no better way to earn someone's trust in your intentions than by listening attentively to what that person has to say — call it active listening. When you listen to others first before communicating your own perspective, you increase the chance they will reciprocate and listen to you. You also learn how to present your message in order to obtain the best possible understanding, appreciation, and agreement.

Active listening shows you care.

Caring. When you take the time to listen to another perspective, you send a most important message that you care. And when people believe you care about them, they will care about what you tell them. They trust you will look out for them when applying your knowledge, skills, and abilities.

Asking questions also communicates caring and builds trust. Not typically generic questions like, "How are you doing?" but deeper, more probing questions. Questions targeting a specific aspect of a person's job send the signal you care about him or her. You're showing genuine interest in what people are doing and how they feel. It's especially powerful when your words reflect actively caring about health and safety. Although extreme, the behavior of the husband in Figure 30 shows he cares, and in the process he builds interpersonal trust.

Candid people avoid prejudice.

Candor. We trust people who are frank and open with us. They get right to the point, whether asking for a favor or giving us behavior-based feedback. Of course, candor does not give you permission to be tactless and inconsiderate as shown in Figure 31. And when candid individuals don't know an answer to our question, they don't ignore us. They tell us outright when they don't know something, and they tell us they'll get back to us later. Then our trust in both their intention and abilities increases when they get back to us soon with an answer.

"It all happened after he took a course at work on behavior-based safety."

Figure 30. Actively caring behaviors at work can spread to the home.

Figure 31. Candor is inappropriate when inconsiderate.

Candor also requires a lack of prejudice. Candid people are not judgmental. We all tend not to trust people who show a tendency to evaluate or judge others on the basis of some stereotype or preconceived notion. How fair are those people going to be? Even when their prejudice is not directed toward you, are you going to trust their abilities and intentions?

Consistency. We usually trust the intentions of people who confess openly their inability to answer our question. But we expect them to follow through when they say they'll get back to us. What happens when they don't — when their actions are not consistent with their words? How often do we make promises we don't keep? Whether the promise regards a positive reward or negative threat of punishment or discipline, trust decreases when the consequence is not delivered.

Commitment. People who are dependable and reliable are not only consistent, they demonstrate commitment. When you follow through on a promise or pledge to do something, you tell others they can count on you. You can be trusted to talk the talk and walk the walk. Making a commitment and honoring it builds trust in both intention and ability.

Consensus. When a group of people reach consensus about something, the group members are signaling trust in the opinions or recommendations of others. If a decision is reached that leaves out a minority view, there might be active or passive resistance on the part of those who have "lost face" in the process. Without everyone's buy-in, commitment, and involvement, we can't trust the process to come off as expected.

Group consensus develops from candid, consistent, and caring communication.

So how can group consensus be developed? It requires candid, consistent, and caring communication among all members of a discussion or decision-making group. Fran Rees (1997) lists six basic steps to reaching a consensus decision:

Follow six basic steps to build group consensus.

- Set a decision goal. What is the aim or purpose of the consensus-building exercise?

- Spell out the criteria needed to make the group decision acceptable. What qualities or characteristics of the decision are needed to satisfy the goal? One criterion might be to stay within budget constraints.

- Gather information. What information is useful for making the decision? Where is this information and who can provide it?

- Brainstorm possible options. Does everyone understand each option and its ramifications? Has everyone had a chance to voice a personal opinion?

- Evaluate the options against the group's criteria. Which solutions meet the "must have" criteria? Which options meet the "nice but not necessary" criteria? Can certain options be combined to meet both criteria?

- Make the final decision as a team. Which option or combination of options best meets all of the "necessary" criteria and most of the "desirable" criteria? Who has reservations and why? How can individual skepticism be resolved? Can everyone support the most popular option? What can be altered in the most popular action plan to attract unanimous support and ownership? I'm sure you see that there's no quick fix to reaching consensus.

Character. A person of "character" is considered honest, ethical, and principled. All of the characteristics discussed here to describe a trusting culture are practiced by a person of character. I'd like to add a few additional qualities:

People with character invite behavior-based feedback.

First, individuals with character admit vulnerability. They realize they aren't perfect and need behavioral feedback from others. By actively listening to others and observing their behaviors, individuals with character learn how to improve their own performance.

Having the courage to admit your weaknesses means you're willing to apologize when you've made a mistake and to ask for forgiveness. There is probably no better way to build trust between individuals than to own up to an error that might have affected another person. Of course you should also indicate what you will do better next time, or you should ask for specific advice on how to improve.

People with character admit mistakes and never "back-stab."

The surest way to reduce interpersonal trust is to tell one person about the weakness of another. In this situation it's natural to think, "If he talks that way about her, I wonder what he says about me behind my back." People with character, as defined here, always talk about other people as if they can hear what's being said.

"Back-stabbing" leads to more back-stabbing, and eventually to a work culture of independent people doing their own thing, fearful of making an error, and unreceptive to any kind of performance-based feedback. Key aspects of behavior-based safety — teamwork and interpersonal observation and coaching — are extremely difficult or impossible to implement in such a culture. And within this context, personal responsibility and actively caring for safety will be improbable. Start to build interpersonal trust by making a personal commitment of no back-stabbing.

In Conclusion

The theme of this chapter is intuitively valid and crucial. People who feel a sense of interdependence, trust, and belonging with their coworkers will also feel responsible for their coworkers' health and safety. They will go beyond the call of duty to prevent their coworkers from getting hurt. So it's important to understand belonging and trust, and to establish work conditions and interpersonal relationships that help to build and maintain these person states throughout a workforce.

Effective teamwork cultivates interpersonal belonging. Thus in addition to enabling synergistic achievements, teamwork builds a feeling state that enhances personal responsibility and actively caring for occupational safety. And the behavioral and attitudinal benefits of teamwork are especially powerful when the team's purpose is related to industrial safety and health.

Personal responsibility for safety develops from interpersonal trust and a sense of belonging.

No other factor is more important in determining the success of a behavior-based safety process than interpersonal trust. Lack of trust stifles the cultivation of belonging and interdependency, and prevents interpersonal responsibility for safety. In a group work setting, the success of behavior-based safety depends upon genuine interpersonal communication, whether we're talking about a formal observation and feedback process, informal individual or group coaching, one-to-one recognition and support, incident analysis and correction, ergonomic assessment and intervention, group goal-setting and celebrating, and so on. All of these situations, and many more I could mention, require interpersonal trust to make them work. And this includes trust in ability and trust in intention, and also includes trust between and across all levels of an organization.

Independent and we/they thinking inhibits interpersonal trust.

A "we/they" mentality stifles trust building. People need to appreciate and respect each other's differences, and realize everyone in the system is interdependent and critical for the organization's synergistic success. Building the kind of belonging and trust needed to break down independent perspectives and we/they barriers is an ongoing and never-ending process.

Building interdependent trust and belonging should be part of the mission statement for every corporate endeavor that involves people. It should influence almost every conversation we have with other coworkers. It is a continuous journey — one that is essential to cultivate an organization of individuals and teams whose personal and shared responsibilities for safety and health are sufficient to achieve a Total Safety Culture.

Teach and Support Safety Self-Management

Chapter 7

Many people consider behavior-based safety to be a one-on-one coaching process whereby one employee (the coach) completes a behavioral checklist while watching another employee work. Then the coach gives the employee feedback by reviewing the "safe" and "at-risk" checks on the list. This observation and feedback process is certainly an important behavior-based safety method, and it can reap great benefits. Safe behaviors will increase and at-risk behaviors will decrease, and workplace injuries will be prevented. Plus, interpersonal trust and group belonging will increase throughout the culture, as workers experience interdependency with regard to personal health and safety.

If implemented correctly behavior-based accountability increases personal responsibility for safety.

A one-on-one coaching process starts with adding a behavior-based accountability system. Then, if implemented correctly (with the principles explained in this book), this increased accountability develops interpersonal responsibility. It's important to realize, however, that behavior-based safety is more than a one-on-one coaching process.

The principles of behavior-based safety, as reviewed in Chapter 1, should be used to guide the development and evaluation of numerous approaches to injury prevention, from incident analysis and corrective-action planning to team building and management systems improvement. And the principles are applicable to many domains other than safety. For example, before focusing my research and scholarship on injury prevention 25 years ago, I applied the behavior analysis principles reviewed in Chapter 1 to improve classroom learning in elementary schools, to reduce theft from newspaper racks, vending machines, and department stores, to manage behavior more effectively and humanely in maximum security prisons, to improve the performance of both beginning and advanced tennis players, and to decrease littering and increase recycling, carpooling, and energy and water conservation throughout communities.

Behavior-based principles and procedures are applicable in numerous domains.

My first behavior-based book (Geller, Winett, & Everett, 1982) reviewed a decade of research my colleagues, students, and I conducted to develop large-scale behavior-based interventions to promote environmental protection. In that case, our challenge was to design an accountability system that motivated people to do various things to protect the environment while also feeling greater responsibility for environmental preservation.

My point here is that behavior-based safety is more than the one-on-one coaching process popularized today by major consulting firms. In fact, I hope you see this book as testimony to the breadth of behavior-based principles. Actually, it's useful to consider the application limitations of a one-on-one coaching process. While a powerful approach to improving behavior and attitude, one-on-one safety coaching requires the presence of another person (the coach), as well as a situation in which critical behaviors are likely to occur. What about those numerous times, like the situation depicted on the cover of this book, when people are working alone? And what about those important safety-related behaviors that occur infrequently and are not likely to be occurring during a scheduled observation and feedback session? Figure 32 illustrates such a situation.

The worker in Figure 32 is performing a critically important task. The various components of this task, from body positioning to using certain personal protective-equipment, are inconvenient or somewhat uncomfortable. And, the safe procedures are not the most efficient. Why is this worker following safe operating procedures? He's working

Figure 32. Critically important safe behaviors are relatively infrequent and often performed alone.

alone, so there's no one around to hold him accountable. He could take a short cut, and get to the break room faster. Someone could hold him accountable for completing the lock-out procedure by checking out the equipment later, but no one can know how this lone worker locked out the equipment. And the probability of incurring an injury while locking out equipment with the wrong body language must be minuscule.

So what motivates the worker in Figure 32 to follow a certain list of lock-out procedures? He's holding himself accountable, which reflects personal responsibility. How did this person get to that level of commitment to safety? And how can an organization help others hold themselves accountable for going out of their way for safety when working alone? That's the theme of this chapter. Interpersonal observation and feedback are not relevant, but principles from behavior-based safety are applicable. In fact, the very essence of behavior-based safety is used for the process described here. I call it safety self-management.

What is Safety Self-Management?

A potent approach to facilitating a transition from safety accountability to personal responsibility for safety is to teach self-management principles throughout your workforce. The method and tools of effective self-management have been derived from behavioral science research and are perfectly consistent with the principles of behavior-based safety. In fact, I've saved this topic for the last chapter, because discussing the techniques of safety self-management will involve a review of most of the key concepts and procedures covered in earlier chapters. Consider first that teaching and supporting safety self-management gives safety the status it deserves. This helps to increase people's sense of personal importance when they perform safety-related tasks (the theme of Chapter 5).

The tools and techniques of effective self-management are derived from behavior-based principles.

Self-management techniques are clearly relevant for situations beyond safety. In fact, the procedures and tools of this behavior-based intervention process were developed and evaluated by behavioral scientists interested in helping people with the variety of personal adjustment issues seen in the clinical psychology clinic. Few if any of these problems relate to safety. For example, a popular and comprehensive review of self-management technology first published in 1970 and now in its sixth edition (Watson & Tharp, 1993) never mentions a safety-related behavior. Personal improvement domains addressed thoroughly by Watson and Tharp with self-management techniques include anxiety and stress, weight loss and overeating, studying and time management, depression and low self-esteem, smoking, drinking, and drugs, exercise and athletics, and relations with others (including social anxieties, social skills, and dating).

Self-management techniques are applicable for self-improvement in numerous areas.

Although specific therapy procedures differ considerably across the various personal adjustment areas targeted with self-management technology, the basic principles are the same and parallel those used in behavior-based safety (as reviewed in Chapter 1). Therefore, it's not difficult to apply self-management principles to safety related problems. And the best news is that over 30 years of rigorous evaluative research has shown that when these basic principles of behavior-based psychology are applied correctly to the variety of personal improvement areas listed above, the self-management works (as reviewed by Watson & Tharp, 1993). So teaching and supporting safety self-management in your company can reap benefits beyond building safety responsibility. But again, by starting with safety you are boosting the importance of safety-related behavior and safety self-management. You're empowering people to actively care more effectively and consistently for the safety of others (the theme of Chapter 4).

From Outside to Inside Direction

Watson and Tharp (1993) emphasize that all behavior passes through three sequential stages: a) control by others, b) control by self, and c) automatization. In this book, I've referred to control by others as "accountability" and control by self as "responsibility." Automatization occurs when behavior becomes a habit. If the habit is good, appropriate, or safe, we say the individual is "unconsciously competent." If the habit is bad, inappropriate, or unsafe, the person is "unconsciously incompetent."

Therefore, our behavior and the behavior we see from others are in one of these stages of development. It is following the directions of others, perhaps according to a step-by-step accountability system; it is self-directed according to one's personal responsibility; or it can be habitual or automatic — cued by environmental stimuli but done without conscious judgment or self regulation.

All behavior is other-directed, self-directed, or automatic.

Other-directed behavior is at risk when the instructions or the example followed is at risk. Self-directed behavior is at risk when self-regulation doesn't match the safe directive, as in a calculated risk (Chapter 3). Automatic behaviors or habits are at risk when the performer is unaware of a mismatch between the safe directive and the current behavior, as in most human error. Thus, safety self-management is essentially two fold. It is a process of matching self-directed behavior to a safety standard, and bringing at-risk habits into conscious awareness or the self-regulated stage so a mismatch between safe and at risk can be realized and adjusted.

From Outside to Inside Motivation

To appreciate the three stages of behavioral development and the execution of safety self-management, let's return to Figure 32. This individual is in the self-regulation stage, but he initially learned the safe steps of power lockout from an outside source. At first he probably followed a step-by-step behavioral checklist, which was perhaps explained during a power-lockout training session. Maybe a supervisor or coworker watched this individual complete his first power lockout procedure and followed his sequential actions with a behavioral checklist. Then in a one-on-one feedback session, the worker was held accountable for his other-directed behavior. Through behavior-based feedback he matched his new sequence of behaviors with the safety standard.

Now the worker is following an internal checklist as he self-directs a safe power lockout process. Why does he follow the inconvenient safe process? Well, first he knows what to do. He was taught well. But knowing what to do doesn't always lead to proper execution. What motivates this individual to be responsible — to match his self-directed behavior with the safety standards he learned previously? In other words, what is he saying to himself to justify his special commitment to safety?

When people feel empowered they are motivated to use self-management techniques to improve.

Much of this book has been about establishing the kind of circumstances and context in which safety responsibility or self-regulation can develop. In Chapter 2, I explained why external negative controls do not cultivate self-directed accountability, and in Chapter 3, I discussed ways to promote empowerment as a feeling state indistinguishable from personal responsibility. When people believe a) they have the ability to make a difference (self-efficacy), b) they are in control of the factors needed to be successful (personal control), and c) they expect to achieve the best (optimism), they are truly empowered. They are motivated to use self-management techniques to continuously improve. They can apply a behavior-based self-management strategy to increase their awareness of mismatches between certain safety standards and their other-directed, self-directed, and automatic behaviors. They increase their self-motivation to follow the safety standards even when such behavior is less efficient, inconvenient, or a bit uncomfortable.

The Basics

Before turning to a discussion of seven self-management strategies for increasing safety awareness and motivating safety-related behaviors, let's review some key points I've covered so far that determine the success of safety self-management. First, it's important to appreciate the importance of empowerment, including perceptions of self-efficacy, personal control, and optimism. Research has shown quite convincingly that individuals who feel empowered to use self-management techniques are going to be more successful. This means they believe they have sufficient knowledge, resources, and support to apply self-management successfully. They also believe the behavior-based techniques will work, and they want them to work. In other words, they are optimistic and they appreciate the significance of improving the self-targeted behaviors. Recall the theme of Chapter 5 — help people feel important by giving them opportunities to choose methods for accomplishing important tasks.

When people feel important they are more likely to use self-management techniques.

Safety self-management is all about personal choice. Individuals choose personal behaviors to improve. They select a plan for monitoring the target behaviors and then design an intervention process to motivate self-improvement. They track their results, assess the degree of improvement, and then decide whether the intervention should be discontinued or refined. They could decide to implement a completely different intervention, or they might be satisfied with the behavioral improvement and decide to target another behavior for safety self-management.

Self-management involves the four steps of DO IT.

If those self-management decisions remind you of the behavior-based safety DO IT process reviewed in Chapter 1, I'm pleased. The basic steps of safety self-management parallel the steps of DO IT exactly. You **D**efine a target behavior; **O**bserve and monitor the target behavior; **I**ntervene with an action plan to improve the target behavior; and then **T**est your results to assess the degree of self-improvement and decide what to do next. Unlike the typical behavior-based safety process, however, self-management is designed, implemented, and evaluated by and for the same individual. It is personal and self-focused, and therefore some of the procedures are different than an interdependent group process, as I explain in the next section.

Activators direct and consequences motivate.

The intervention procedures for self-management are different than the procedures used to change team behaviors in an interdependent context, but the interventions are still developed from the basic three-term contingency or activator - behavior - consequences (ABC) principle. As reviewed in Chapter 1, activators direct behavior and consequences motivate behavior. And as illustrated in Figure 33, the direction provided by an activator is likely to be followed when it's backed by a consequence that is soon, certain, and significant. In the next section, I'll discuss ways to arrange activators and consequences to self-direct and self-motivate personal behavior.

Although self-management techniques are self-planned, self-implemented, and self-evaluated, other people can serve an invaluable support service. For example, you might choose a family member or coworker to be an activator — to re-mind you about a particular behavior. Or you might ask a friend to dispense a particular reward when you deserve it. If you think about it, the ultimate motivation for most self-improvement programs, including safety self-management, is to please someone else — a family member, a friend, a supervisor, a coworker. And research shows conclusively that a person's self-management process can be benefited greatly by appropriate social attention and appreciation. Therefore, the recommendations in Chapter 6 for increasing belonging and interpersonal trust are relevant here. People are more motivated to improve for coworkers they like and trust, and team members can offer invaluable advice and

CHAPTER 7 — TEACH AND SUPPORT SAFETY SELF-MANAGEMENT

Figure 33. Activators work when backed by a consequence.

encouragement for an individual's safety self-management project. But why should someone choose to engage in a self-management improvement process for safety. What can we say to others to convince them to participate in safety self-management? Let's consider a few answers to these important questions.

Getting Started in Safety Self-Management

The first step of safety self-management is to recognize a problem or a need for improvement. And that's easier said than done, especially in the realm of safety. Most of us believe we're safe enough. If we weren't we'd be getting hurt — right? That sounds like good common sense, but we know it's not true. From an individual perspective, the self-statement "It's not going to happen to me" makes sense, and is verified everyday. I think teachers and consultants who ask us to stop believing or saying this phrase are making a mistake. You can't discredit a self-statement that is verified consistently by personal experience. If a person really thought it was going to happen to them, they would have a good excuse to stay home — right?

From a group or workplace perspective, the statement "It's not going to happen" is not true. And this can be verified by industrial injury reports. So it is fair to say "It's going to happen to someone." Then ask the relevant follow-up question "Do you care?" I've never heard "No" given as an answer to this question, meaning most people want to help reduce the probability of coworkers getting hurt. But they might not know how to "actively care."

One sure way to actively care is to always demonstrate the safe way, because if you don't, you are inadvertently teaching others to take the same calculated risks and put themselves in some degree of danger. And if everyone takes the same risk, someday someone will get hurt. From an individualistic viewpoint, the probability of an injury might be small, even minuscule, but from a big picture, collective outlook the probability approaches certainty.

At the individual level an injury is improbable, but at the organizational level an injury is almost certain.

Now we have a reasonable self-statement to use as justification for engaging in safety self-management. We need to practice the safety standards consistently, because people learn by watching us. And if we demonstrate an at-risk shortcut, we could be teaching others to make an at-risk decision. And they will teach others, and the others will teach others, and so on until at-risk behavior becomes the norm. With an entire work force exposing themselves daily to a particular risk, someone will eventually get hurt, regardless of how small the probability of injury is at the individual level.

Thus, people who care about others getting injured and who are "emotionally intelligent" enough to take a broad and long-term view of their current behavior, want to practice safe operating procedures. Of course the more belonging and interpersonal trust we feel toward our coworkers, the more we care about them avoiding personal injury and the greater is our internal justification or responsibility to always demonstrate the safe way of doing something.

Sometimes when I present this rationale for engaging in safety self-management at workshops and seminars, a participant will argue that my reasoning doesn't hold up in situations when people are working alone. "It makes sense that people learn our at-risk behavior when they can see us," they retort, "But when we're working by ourselves, there's no risk of observational learning. So why not take our calculated risks when working alone, and save our inefficient safe procedures for times when others might see us?"

One reaction to the response discrimination proposal is that you can never be sure another person won't see you and learn from your risk taking. However, there are times when the probability of observational learning is very low — then what? This reminds me of the excuse I hear from people about not always using their vehicle turn signals. They tell me they discriminate appropriately, and only use their turn signal when other drivers can see their signal. "That's interesting," I reply, "You waste valuable cognitive processing or brain power to decide whether or not the availability of other vehicles calls for turn-signal use, instead of just using your blinker to signal every turn and thus put the behavior in automatic mode. Why not become 'unconsciously competent' with regard to turn-signal use, and save your conscious decision making for dealing with the periodic uncertainties of driving?"

Chapter 7 — Teach and Support Safety Self-Management

Figure 34. Think globally but act locally.

Simple behaviors like turn-signal use should become a habit.

Perhaps you recognize that the argument just given for always signaling a turn or lane change relates to the three stages of behavioral development — other-directed, self-directed, and automatic. It makes sense to get as many safety-related behaviors as possible into the automatic mode. Then we can use the self-directed stage to increase personal control over new other-directed behaviors, or to bring current self-directed or automatic behaviors more in line with ideal safety standards.

Automatic behavior is under environmental control.

Note that automatic behavior is under environmental control rather than self-control. In other words, you don't need to think about the habit; it is just triggered by an environmental event. It's certainly possible, however, that your automatic behavior does not meet ideal safety standards. In this case, self-management is needed to retrieve the at-risk habit and bring it to the self-directed stage. Here you can refine or alter the behavior to meet your safety standard. Perhaps new environmental stimuli need to trigger an underused safety procedure, or old stimuli need to stop activating a prior habitual at-risk behavior. At any rate, safety self-management is a process for self-regulating new and old behaviors in order to meet certain safety standards.

Some of your improved behaviors will become automatic, as in "unconsciously competent," while others are sufficiently complex and infrequent (as with power-lockout behaviors) to always require self-direction or "conscious competence." But given the limitation of our mental capacity to self-direct, it's wise to relegate as many simple safe behaviors as possible to the automatic mode (like use of vehicle safety belts and turn signals). Then you have cognitive capacity to adjust your more complex behaviors to meet your safety standards. But what are your safety standards? Are your safety standards ideal? Is safety a value or a priority in your life?

Safety as a Value

I often discuss the need to envision safety as a value rather than a priority. Priorities shift depending on current needs and contingencies. The number one priority today might not be the number one priority tomorrow. Many dynamic factors will move other problems, issues, goals, or tasks to the top of the demand hierarchy. Change is a constant. And leading change or adjusting to it requires flexibility and strategic modification of priorities.

Priorities change, values remain.

Values are more constant than priorities. For individuals, values represent profound internal beliefs or attitudes that establish a context from which we evaluate past behavior and plan future behavior. Likewise, the values of an organization are defined in the mission statement and provide direction for short- and long-term action plans.

Given this definition of "value," it becomes obvious why it's better to talk about safety as a value rather than a priority. Priorities are compromised to make way for other priorities. Values are rarely compromised. They serve as the standard against which we judge the appropriateness of our behaviors. When our actions are inconsistent with our values, we willingly make appropriate adjustments to align behavior with value. Thus, if safety is considered a value, the safest way of doing something becomes the standard against which ongoing work practices are evaluated; and if an inconsistency is pointed out, behavior is willingly changed.

Our values are standards against which we compare our behaviors.

Do you hold safety as a core value? Do you believe the personal safety of an employee is more important than productivity? Do you believe safety is so important that it should be linked to every priority? Regardless of the work demands of the day, should personal safety come first?

From my experience, it's not difficult to get people to respond "yes" to each of these questions. People like to think they believe in safety. Now if people claim safety as a value, the next step is to remind them of the meaning of a value, as defined above. Values determine the ideal standard against which we compare our behaviors. Hence, if our other-directed, self-directed, or automatic behaviors don't match the standard set by our values, self-management is needed.

At a recent seminar on behavior-based safety, a group of top executives expressed concern that their line workers would not implement a behavior-based safety process. They were particularly worried that the union would resist participating in an interpersonal observation and feedback process. I began discussing how to build interpersonal trust and how to start a behavior-based feedback process that is not overwhelming. During this discussion one of the executives indicated that their union has declared safety a core value. That important statement caused me to alter my advice on how to "sell" line workers on a DO IT process of behavioral observation and feedback.

Chapter 7 — Teach and Support Safety Self-Management

If safety is declared a core value, then the need to adjust behaviors inconsistent with the ideal is obvious. And, the rationale for getting started in a safety self-management process is clear. People strive to maintain consistency between their values and their behaviors (Cialdini, 1993), so they should be naturally motivated to use self-management techniques to assure the consistency.

> *People strive to maintain consistency between their values and their behaviors.*

You might need to remind people that safety is a core value, and that safety self-management is an effective process for maintaining consistency between behavior and value. You can only get a rational argument from someone who's not willing to answer "yes" to the value questions posed above, and I bet you won't find many people willing to admit to such irresponsibility.

Incidentally, when you observe an inconsistency between a value and a behavior, it's not necessary to apply distasteful punishment procedures (as covered in Chapter 2). Instead, point out the inconsistency and expect a change consistent with the value. This illustrates the Principle of Consistency — a powerful determinant of behavior and attitude change. In fact, I think achieving consistency between behavior and the right value is key to being responsible to "The Man in the Glass," the inspirational poem by Harvey Holland Upchurch reproduced in Figure 35.

The Man in the Glass
By Harry Holland Upchurch

When you get what you want in your struggle for self
And the world makes you king for a day,
Just go to the mirror and look at yourself
And see what that man has to say.

For it isn't your father or mother or wife
Whose judgment upon you must pass.
The fellow whose verdict counts most in your life
Is the one staring back from the glass.

Some people may call you a straight shooting chum
And call you a wonderful guy,
But the man in the glass says you're only a bum
If you can't look him straight in the eye.

He's the fellow to please, never mind all the rest,
For he's with you clear to the end,
And you have passed your most dangerous test
If the man in the glass is your friend.

You may face the whole world down the pathway of life
And get pats on the back when you pass,
But your final reward will be heartache and strife
If you've cheated the man in the glass.

Figure 35. We are responsible for matching our behavior with our core values.

An Illustrative Anecdote

What would your reaction be if you received a phone call from a police officer regarding your child? Your heart would undoubtedly pound furiously with fear of hearing some terrible news. If the call was to inform you that your 16-year-old would be held overnight in jail for consuming alcohol if you didn't go to the station, how would you react? Would your fear turn to anger? Would you begin planning a series of punitive consequences (referred to as "discipline" in industry) so as to make certain this teenager would never consume alcohol again before age 21? Five parents got such a phone call from the local police department in December, 1996 (on Friday the 13th). Their daughters, all 16 years old and varsity athletes, were stopped, searched, handcuffed, and arrested for underage drinking. One of those teenagers was my daughter.

My daughter's behavior was inconsistent with values taught through experience.

Inconsistent behavior. My wife Carol took the phone call from the police officer, and drove to the station. Yes, this was the teenager I've written about many times to illustrate youth involvement in safety efforts. At age three and a half she held up a large sign in the window of my car with the message "Please Buckle Up — I Care." In the fourth grade, she gave a speech about this "flashing" experience for safety and won the "Best Speech of the Year" award. At age 14 she helped me demonstrate to the media that young teens can too easily purchase cigarettes by attempting to buy cigarettes in 20 different stores. [She was only turned down twice]. And she has also helped my students and me in our government-funded research designed to develop interventions for preventing alcohol abuse and alcohol-impaired driving.

Now my daughter was caught doing something completely inconsistent with the internal value that should have developed from her personal involvement in safety and health promotion. I was reassured she had the right values when she sobbed uncontrollably after seeing her mother. "I'm so embarrassed, Mom" she said, "This is like a bad dream; this is not me." "Right," said Carol, "This is not you."

What consequences? What punitive consequences are appropriate for this situation? Under-aged drinking is surely a tragic public health problem, and the thought that such behavior could lead to alcohol-impaired driving is downright scary. So what should a parent do? It was interesting, and predictable, to learn what the parents of these five teens did.

When we don't own up to our failures, we don't experience inconsistency between our values and our inappropriate behaviors.

As you might have guessed, the common reaction was to exert additional punitive controls. At a minimum, all of the girls were "grounded" over the Christmas holidays. One lost her car privileges, which forced her to take the bus to school (humiliating for this teen). Another girl was forbidden to interact socially with any of the other four cohorts. Incidentally, this individual had claimed she was not drinking the alcohol — a lie she created to avoid additional punishment. There, the threat of external control motivated another undesirable behavior. When we do not own up to our failures, we don't experience maximum inconsistency between our behavior and our values. Then we don't feel obligated to change our behavior.

According to The Consistency Principle, my wife had the best reaction in this situation. She reinforced the inconsistency between our daughter's behavior and her values. Carol actively listened to her description of the humiliating experience of being searched and handcuffed and escorted to the police station. The police officers appeared unimpressed with our daughter's report that she had assisted my research projects to prevent drunk driving. By responding with empathy, mom enabled her daughter to own the inconsistency between her behavior and her values. This maximized the probability of future values-driven behavior.

Chapter 7 — Teach and Support Safety Self-Management

Carol and I are thankful for the external controls that threaten humiliating penalties for noncompliance with drug-use laws. And we are grateful our daughter experienced the punitive consequences, and appeared in court to state remorse. But we also realize these external controls are not sufficient to maintain safe behavior. The perceived improbability of getting caught again cannot compete with the daily influence of peers tantalized by "forbidden fruits." We can only hope pressure to avoid inconsistencies between behavior and value will be more powerful than pressures to conform with the risky behaviors of others.

The connection. My personal anecdote took our discussion beyond safety and health issues in the workplace, but I hope the connection between a family's experience with under-aged alcohol consumption and values-driven safety is clear. An external threat is often most convenient and expedient, but it has clear disadvantages if you're attempting to make safety a value. In some cases, severe threats can actually make the undesirable behavior seem desirable. In situations with exclusive top-down controls and minimal bottom-up empowerment, getting away with noncompliance is reinforced with outside peer support and inside feelings of individual freedom (as discussed earlier in Chapter 3).

Values-driven safety cannot be dictated.

We need to understand and believe values-driven safety cannot be dictated. Values can be developed from outside sources that encourage voluntary participation in activities representing or supporting a particular value. We internalize the principles and lessons we choose to experience, but we are apt to resist those principles and lessons we feel are forced upon us. Obviously, it's a difficult but important challenge to provide just enough outside accountability to make a program or process appear worthwhile and inviting without inhibiting feelings of personal responsibility. This book has attempted to provide answers to this crucial issue.

The new lesson I want you to get from this story is that the Principle of Consistency offers a powerful reason for engaging in safety self-management. If we hold safety as a core value then it is incumbent on each of us to hold ourselves accountable for matching our everyday behaviors with ideal safety standards. Let's see how safety self-management can help us do that.

The Techniques of Safety Self-Management

As with the DO IT process, safety self-management starts with defining one or more target behaviors that need improvement. These might be safe behaviors you need to perform more often to meet a safety standard, or they could be at-risk behaviors you need to perform less frequently to assure consistency with holding safety as a value. You probably know where your weaknesses are with regard to complying with ideal safety procedures. However, when starting out, you should not develop a long list of behaviors to change. Start with a few target behaviors that need improvement and are critical to injury reduction.

Start with a few critical behaviors that need improvement.

Don't rely completely on common sense to select a critical behavior to work on. If lots of your safety-related behaviors need improvement, then start with behaviors that are most critical with regard to preventing serious injury to yourself or someone else. Start with behaviors that are relatively easy to change. Then when you experience success, you'll be ready to add more critical behaviors to your safety improvement list. For example, if you currently don't buckle your safety belt consistently in every vehicle you drive, from a personal automobile to the company fork-lift truck, this behavior should be on your list. Safety-belt use is among the most convenient and protective behaviors you can perform, and your use or non-use of a safety belt can influence a lot of observers.

Consider the use of PPE critical. Actually, all use of relevant personal protective equipment (PPE) is critical and relatively easy to perform. So if you currently do not use all PPE consistently, you should consider working on these behaviors. Don't hesitate to ask advice from your safety director or anyone else whose job is to manage safety in your workplace. This individual can help you consult injury records and near-miss reports to see which behaviors contributed to these incidents. And it might be appropriate to consult an ergonomics specialist regarding particular repetitive behaviors in your work setting that need adjustment to reduce the probability of a cumulative trauma disorder.

Self-Observing and Recording

Self-knowledge is key to successful self-management. Most of us believe we understand ourselves, and we place great faith in our ability to remember our own behaviors. So, you may feel it's not necessary to systematically observe your own safety-related behaviors. Unfortunately, our memories of behavior are much less accurate than we think. Research indicates that people have much more confidence in their memories than they should. In one weight-loss study, for example, people were asked to write down everything they had eaten over the last two days. According to these lists, these people were not overeating. But when these people ate only the food they had reported eating, all of them began to lose weight.

Sample a variety of safety-related behaviors before selecting your target. In order to improve your safety-related work practices, you need to obtain a record of your safe and at-risk behaviors. In fact, before selecting a few critical safe behaviors to work on, you might need to sample several safety-related behaviors in various situations. Your memory of what behaviors need improvement might be as faulty as the memories of the participants in the weight-loss research. Of course, when you begin self-observing and recording certain behaviors, you'll pay more attention to these behaviors. So your record will likely be an overestimation of your personal safety. That's good of course, and shows the benefit of bringing behavior to the self-directed stage for evaluation and self-improvement. This observing and recording process is facilitated with the right kind of behavioral checklist.

Self-observing and recording will affect self-improvement. *Developing a self-observation checklist.* A good self-observation checklist (SOC) has a number of qualities. The most effective SOCs are the following:

- *portable and accessible* whenever the behavior(s) occur,
- *easy and convenient* to use,
- *noticeable*, so they can serve as an activator for making observations,
- *records of both safe and at-risk behaviors*, allowing for a computation of percentage of safe behaviors.

Keep your self-observation checklist simple. An SOC with these characteristics will benefit self-management by making the recording process easy and convenient. The SOC depicted in Figure 36 has all of these qualities, and is simple to use on a daily basis. Space is provided for five target behaviors. Every time an opportunity for a particular behavior occurs, you merely judge whether the behavior was safe or at-risk and mark the checklist accordingly. The observation procedures are exactly as I have detailed elsewhere for interpersonal coaching

CHAPTER 7 — TEACH AND SUPPORT SAFETY SELF-MANAGEMENT

(Geller, 1997c), except here the observers hold themselves accountable for certain behaviors instead of another person.

Charting progress affects self-improvement.

Multiple opportunities to perform each target behavior can be tabulated on the SOC in Figure 36. At the end of a day or week, the number of safe and at-risk occurrences of each behavior can be totaled to calculate a percent safe score (number of safe occurrences/number of safe + at-risk occurrences) for each behavior. And an overall percent safe score can be calculated from the totals. These percent safe scores can be charted on a graph and used to assess daily or weekly progress. Obviously, the mere recording and charting of percent safe scores will lead to significant improvement because the process is a means of holding yourself accountable. But as I discuss below, there are additional intervention strategies you can use to make success easier and greater.

Record occurrences of checklist behaviors as soon as possible after they occur.

Using a self-observation checklist. Delays in recording occurrences of the target behavior(s) can weaken the improvement process. As discussed above, our memories for specific behaviors become less accurate over time. So it's not a good idea to wait until the end of the day and then try to remember how many times you engaged in a particular target behavior. Your count will not be accurate.

Obviously, the most accurate accountability occurs when you record occurrences of behavior immediately. Try to anticipate problems you might have making immediate recordings of your behavior and look for ways to deal with them. For example, if the behavior is something you engage in quite frequently, recording on the SOC can be cumbersome and tedious. In these cases, a wrist counter like those used for golf would be quite useful.

Self-Observation Checklist

Target Behaviors:	Safe	At-Risk	% Safe
1) _____			
2) _____			
3) _____			
4) _____			
5) _____			

Comments: Totals _____

_____   **Overall % Safe Behaviors** _____
_____

Figure 36. A self-observation checklist can be simple.

Note activators and consequences related to each checklist behavior.

Taking note of specific activators and consequences of your safety-related behaviors can provide valuable insight into the events that control your other-directed and habitual behaviors. Record these observations in the comment section of your SOC (see Figure 36). This information can be invaluable when developing a self-management plan. For example, information about activators that set the occasion for at-risk behaviors tells you what situations require special attention, and might suggest ways to manage certain activators (a self-management strategy discussed in the next section). And, information related to the various consequences motivating specific safe and at-risk behaviors could help you devise a plan for making it more rewarding to choose the safe behavior and less rewarding to choose the at-risk behavior.

Continue your self-observing and recording for several weeks.

The fewer the critical behaviors targeted for self-observing and recording, the faster and greater the impact. Thus, targeting a few behaviors at a time usually improves success. Also, the greater the frequency of self-observing and recording per behavior, the greater the beneficial impact. The length of time for making self-observations of a specific target behavior will vary across situations and behaviors. In most cases, however, self-observations should continue for at least three to four weeks.

In summary, the process of self-observing and recording critical safe behaviors versus at-risk behaviors has a number of benefits, including:

- *increasing awareness* of personal safety performance,
- *identifying activators* that influence other-directed and automatic behaviors,
- *identifying consequences* that motivate certain safe and at-risk behaviors,
- *providing information* that can be used to develop an action plan for improving behavior and tracking progress toward accomplishing safety goals.

Intervening for Self-Improvement

The next phase of DO IT and safety self-management is intervention. This is when you derive a plan to improve certain critical behaviors that do not meet your safety standards. This essentially means safe behaviors need to occur more often and/or at-risk behaviors should occur less often. You might need to conduct the observing and recording stage several times on various behaviors before pinpointing critical behaviors requiring your intervention attention. As I've indicated earlier, the observing and recording process alone will bring other-directed and habitual behaviors under self-control and this might be sufficient for desired self-improvement.

Select a few behaviors to target and then develop a self-management plan.

On the other hand, your self-observing and recording processes will likely point out ways to manage the situation, in particular activators and consequences, in order to improve safety performance. With this process you will pinpoint behaviors that need self-improvement. Select a "manageable number" of these behaviors to focus on first. While there is no specific prescription for the optimal number of behaviors to address at once, it is better to start with a few behaviors and build. Once the target behavior(s) are selected, a self-improvement plan can be developed. Let's turn to a discussion of the behavior change strategies you can choose from when designing a self-management intervention.

Chapter 7 — Teach and Support Safety Self-Management

Activator Management

Activator management involves identifying environment, behavior, and person factors that precede the occurrence of safe and at-risk behaviors. Strategies are then employed to eliminate activators that precede at-risk behaviors and add activators that can increase the probability of safe behaviors. For example, simple reminder messages, strategically placed, can be quite useful. One such approach is illustrated in Figure 37.

Discovering activators that use safe or at-risk behaviors is an important first step in developing an effective self-management plan. And these activators can be identified by referring to the baseline information collected during your self-observations. The comment section on your self-observation checklist should include useful insight about activators (see Figure 36).

Posters, signs, and reminders are perhaps the most popular activators for safety. Some bear a general message — "Safety First;" others refer to specific behaviors — "Hard Hats Required." Some signs request the occurrence of behavior — "Wear Ear Protection in This Area," others prompt you to avoid a certain behavior — "No Running." Some signs imply consequences — "To Avoid Eye Injuries, Wear Safety Glasses at All Times," others do not — "Wear Safety Goggles."

Only a small portion of the many activators we see each day actually influences our behavior. Many of us are in a state of "information overload." It seems everyone wants

Figure 37. Activators can be added to the environment for self-management.

to be an outside director of our behavior. Thus, it is better to use a few powerful safety activators than to add more activators to a system already overloaded with information.

However, the self-observation process draws our attention to relevant activators. This is good. We might become aware of an existing safety activator, and use this reminder to bring at-risk habits under safe self-directed control.

If you have an occasion to design your own safety activator for yourself or others, consider the following guidelines:

- specify the behavior
- maintain salience with novelty
- vary the message over time
- activate close to the response opportunity
- imply consequences
- involve other employees

The most powerful activators indicate positive consequences gained or negative consequences avoided by a specified behavior.

If you want an activator to motivate action, you need to specify the desired behavior and imply consequences. The most powerful activators make the observer aware of positive consequences to gain or negative consequences to avoid by performing the target behavior. Sometimes a picture is all you need. Then you write your own script to go with the picture.

Consider a picture like that illustrated in Figure 38, for example. By depicting a possible negative consequence from nonuse of personal protective equipment, that picture can be worth 1000 words. The picture also shows the safe behavior needed to avoid injury. When observers identify with the character in this illustration, they add internal scripts or self-statement to make the activator more personal and more powerful.

Self-Statements

Self-statements are activators. They can cue certain safe or at-risk behaviors. When getting in your vehicle you might tell yourself, "I'm only going a few blocks, so I really don't need to put on my safety belt." Then you probably won't buckle up. But if you say to yourself, "I need to buckle up to set a safe example. My behavior could influence anyone who sees me, especially a family member, and I don't want anyone to think I'm irresponsible about personal safety," you're likely to buckle up. Notice that this safety script includes specific self-instructions, while also expressing a belief about example-setting which provides strong rationale for performing the behavior.

Self-directed messages come in three basic forms:

- self-instruction
- belief
- interpretation

Self-instruction. Self-instruction can be obvious such as a clear inner voice saying, "I need to use my legs when lifting this box." Or the instructions might be suppressed, and occur only as a "faded voice," so weak that only careful attention can reveal them. In this case, your challenge is to turn up the amplifier on this small voice and hear yourself

Figure 38. Activators that show how to avoid negative consequences can be powerful.

reiterating the safe way to do something. In this case self-direction means amplifying your inner voice until the safety-related words are loud and clear. It's a matter of bringing your inner voice into conscious awareness — the self-directed phase of behavior development.

Beliefs. Beliefs are the underlying assumptions that can provide useful rationale or motivating consequences for a certain behavior. They are often not clearly available in our inner speech, but they certainly influence what we say to ourselves. They also determine whether particular self-instructions are loud and clear or weak and ambiguous.

Personal beliefs influence self-statements and their impact on behavior.

It's useful to reflect on your beliefs, and evaluate how they are influencing your self-statements. Do you really believe safety should be a core value, uncompromised by fluctuating priorities? Do you sincerely believe your behavior can influence the behavior of onlookers, and that this makes you responsible to always set the safe example? Do you honestly believe the behavior-based approach to safety is more effective than the more traditional attitudinal perspective?

You can see how your answers to these belief statements will influence the nature and strength of your self-statements. This critical role of belief also explains why it's important to include powerful rationale or intuitive principles with any safety training. If people hear instructions as only directions without personal meaning and significance, their behavior will be only other-directed compliance and will not reach the self-directed stage needed for personal responsibility.

Sometimes it's useful to infer your safety beliefs from your self-observations. "What does my vehicle speeding or calculated shortcut say about my safety beliefs." Perhaps you'll find you're not being entirely truthful when professing, "Safety is one of my core values." As I discussed earlier, people who periodically take calculated risks cannot hold safety as a core value. Realizing an inconsistency between a value you believe is important and a self-observed behavior can be invaluable in motivating the development and implementation of a behavior-based self-management plan.

Personal interpretations determine whether self-management is called for.

Interpretations. What we say to ourselves about the events in our lives influences our beliefs and our self-instructions. And, of course, our beliefs influence our interpretations. In Chapter 3, I discussed the "self-serving bias" as a way to protect our self-esteem by projecting our failure onto outside influences, including other people. And earlier in this chapter, I discussed the need to own up to our blunders and calculated risks in order to recognize inconsistencies between value and behavior, and adjust our behavior accordingly. These are contrary approaches to interpreting the same incident, and will have differential impact on self-statements. A self-serving interpretation implies no need for change, but admitting one's error provides incentive for self-management and self-improvement.

An illustrative experience. On a warm Saturday in November of 1992, my new GEO Tracker was smashed on the driver's side by a large Ford pick-up truck. Yes, I was the driver, and fortunately no one was hurt. Everyone - truck driver, passenger and I - were buckled up. But both the Tracker and truck were severely damaged. The crash occurred at an intersection one block from the house I've called home since August 1972. So I've traversed this intersection safely numerous times. This time was different.

The police officer could not determine any fault in this crash, so no traffic tickets were issued. Each intersection had a stop sign, and both vehicles — truck and Tracker — were stopped before taking off and crashing at very low speeds. It would have been easy for me to interpret this incident as a "freak accident," requiring no adjustment in self-statements or behavior. But I just couldn't do that.

My interpretation of the crash influences my self-instruction.

I remember vividly the crash experience. I've relived it many times, along with my feeling states. The bottom line: I didn't know what hit me. I frankly interpreted the incident as *my* fault. I was hit on the right by a vehicle leaving the intersection at apparently the same time as mine, so legally I could claim the right-of-way. However, that's not a fair assessment. Remember, I didn't know what hit me, so I must not have seen that truck.

Sure I always look several ways at this 5-way intersection. But my interpretation cannot let me off the hook. I may have looked, but I did not see; and that calls for self-management. My habitual intersection behavior was retrieved for self-direction. New self-statements were attached to that driving behavior, and they've been there ever since. Each time I approach this intersection I talk to myself about the necessary stop-and-look behavior.

My vivid images of the crash influence my self-instruction.

The view to the left, the intersection from which the pick-up truck came, is partially blocked by a large building and thus looking takes longer than normal. And although it's rare that a vehicle is there, I remember the time I met one there under unpleasant circumstances. Incidentally, my self-statements at this intersection are facilitated by vivid images of that crash in 1992. In this case, imagery is a powerful activator and motivator of self-statements and self-directed behavior. So it's time to discuss mental imagery as a self-management technique.

Mental Imagery

Mental imagery is using your "mind's eye" to picture situations without actually being there. It's a way to anticipate and prepare for events. It can be used to direct behavior (as an activator) and to motivate behavior (as a consequence). More specifically, for safety self-management you can use imagery to accomplish the following:

- clarify your safety goals
- enhance your motivation to choose the safest behavior
- build your self-efficacy, personal control, or optimism
- rehearse safe acts and actively caring behaviors
- reward yourself for success at self-management

Imagery can direct and motivate behavior.

Your imagery can activate a chain of safe operating procedures, as well as motivate action. The motivation comes from imagining potential consequences following safe versus at-risk behavior. Figure 38 depicts a negative consequence one might visualize to motivate the use of personal protective equipment.

It's often more useful to create a mental picture of positive consequences resulting from your safe actions. By focusing on positive outcomes from safe behaviors you anticipate achievement and the good feelings it brings. This can increase your confidence in being successful, as well as increase your desire to reach your goals. Imagery can also help to clarify what you need to do in order to accomplish your goals.

When using mental imagery for safety self-management:

- see yourself performing the appropriate safe behavior with ease and convenience
- visualize avoiding specific negative consequences with the safe behavior
- imagine feelings of accomplishment following the safe behavior
- take an active rather than passive perspective
- share your imagery with others

When imaging consider the entire activator-behavior-consequence sequence.

It's important to be active in your image. Don't see yourself as a passive observer watching a movie with you in it. Rather, imagine yourself acting within the complete activator-behavior-consequence (ABC) framework. That is, see the activators in the situation which cue the desired behavior. Then visualize yourself actually performing the safe acts, and imagine positive feeling states from setting the safe example and acknowledging safety as a value.

Before imaging the ABC sequence, I find it quite motivating to visualize the negative consequence of getting hurt, as illustrated earlier in Figure 38. When I reach for the skill saw I imagine getting a finger caught in the blade. I imagine the ringing in my ears getting worse after not using hearing protection. And all too often I imagine one of my two daughters in a vehicle crash. Such negative imagery might be too morbid for you, but it sure motivated me to assure my family believed in vehicle safety belts and always used them.

I have also found it useful to share my motivating imagery with others. In fact, one reason personal testimonies of injuries or near misses are powerful motivators is because the listeners can get a mental image.

Personal testimonies activate mental images.

They can readily visualize the speaker in the precarious situation described, especially if the presenter gives a passionate and realistic delivery. Even more motivating is when listeners can put themselves or family members in the situation that caused negative consequences for the speaker. Of course, it's essential to focus on the specific behaviors that can be performed to avoid the injury or near miss discussed.

The lesson here is that it can be useful to share personal experiences in ways that conjure a motivating image. But sometimes the same situation will not evoke the same mental picture for different people. Figure 39 illustrates how imagined consequences can influence ongoing behavior. In this case, the passenger should share her image with the driver. But this might not change the driver's behavior, if his imagery paints a contrary picture. I see a root cause for interpersonal conflict, don't you?

Self-Rewards

In addition to imagining positive consequences following occurrences of a desired target behavior, you can arrange to experience them directly. These are referred to as self-rewards in the self-management literature. Research has shown that individuals who use self-rewards are more likely to sustain their self-management process and maintain self-improvement. Three factors should be considered when using self-rewards:

- selection of the reward
- delivery of the reward
- timing of the reward

Giving yourself the opportunity to do something you like to do can be a powerful self-reward.

Individualize your reward. People like different things, as illustrated in Figure 40. The items in this list were given by Watson and Tharp (1993) as actual self-rewards their clients used in effective self-management projects. The variety of possibilities is obvious. It is important that the rewards be accessible and convenient to use following the behavior, so some potential rewards may not be practical.

Figure 39. Mental images can vary dramatically across people and influence different behaviors.

Notice that all of the examples in Figure 40 are stated as behaviors or activities. That's a nice way to think about rewards. You give yourself an opportunity to do something you really like to do after doing something that is less fun but important for self-improvement. Self-rewarding activities can be internal (as in self-dialogue) or external (as in spending money for food, clothes, or entertainment). Often the strongest self-reward is simply the personal pride associated with the objective and charted accomplishments resulting from self-management.

It's important to specify criteria for obtaining a reward, and to reward successive improvements to the target behavior(s). Rewards should be based on behavior, not affirmation or intention. First, self-observe your behavior in a systematic manner and dispense rewards according to the criteria you define. The reward should come after the target behavior, not before. Rewards should also be administered as soon as possible following the target behavior.

Use shaping principles by self-rewarding small successive steps toward success.

Initially, give yourself rewards to recognize small steps of progress toward the goal. Remember the shaping principles discussed in Chapter 4. Often many approximations to a target need to be rewarded before a goal is reached. After the target behavior improves, reward frequency can be thinned and eventually faded out. However, if self-observations indicate a decrease in the target behavior, reinstatement of the self-reward may be needed.

During the early stages of self-management it's often useful to reward oneself for participating in the process. For example, in one self-management program, the individual rewarded herself by adding $1 to her vacation fund each day she completed her self-observation and recording process. Remember, the most successful self-rewards are readily accessible, individualized, valued, varied, and follow the targeted behavior as immediately as possible.

Goal-Setting

I hope you recall our discussion of goal-setting in Chapter 4. We need to revisit this technique again, because setting personal goals is one of the most powerful ways to self-manage safety. Behavior-based improvement goals should be set high, yet be achievable. They specify expectations for improvement, and provide for the tracking of progress. The acronym SMART was introduced in Chapter 4 as a way to remember the quali-

- ✗ praising oneself
- ✗ taking bubble baths
- ✗ making love
- ✗ going to a movie or play
- ✗ going to the beach
- ✗ mountain climbing
- ✗ spending time at a favorite hobby
- ✗ spending money
- ✗ playing records
- ✗ listening to the radio
- ✗ eating favorite foods
- ✗ going out "on the town"
- ✗ playing sports
- ✗ pampering oneself
- ✗ taking a "fantasy break"
- ✗ window shopping
- ✗ adding money to a vacation fund
- ✗ going out to dinner
- ✗ putting on makeup
- ✗ "doing anything I want to do"
- ✗ going to parties
- ✗ being alone
- ✗ "doing only the things I want to do, all day"
- ✗ goofing off
- ✗ watching TV
- ✗ gardening
- ✗ making long-distance calls
- ✗ buying a present for someone
- ✗ spending extra time with a friend
- ✗ reading mystery stories
- ✗ lip-synching (pretending to be a rock star in front of a mirror)
- ✗ sharing the benefits of self-management with others

Figure 40. Many activities can be used as self-rewards (adapted from Watson & Tharp, 1993).

ties of the most influential goals: Specific, Motivational, Achievable, Recordable, and Trackable.

The goal should specify the behavior(s), the amount of change desired, and the time period in which the goal will be achieved. Too often goals are either unrealistically difficult or too vague to be meaningful and objectively assessed. When goals are realistic and under the individual's perceived control to achieve, they will be achieved. As a result, the individual feels empowered to reach future, more challenging goals.

Goal-setting helps you focus on improving specific critical behaviors. The target behavior(s) should be observed and recorded, and charted on a graph to track progress. Sometimes it's useful to post one's self-management results publicly as a way to gain social support from coworkers. This is especially beneficial when everyone on a work team is working on their own safety self-management project. Then social support becomes a powerful mechanism for rewarding progress, activating additional improvement efforts, and setting expectances for continuous improvement.

Social Support

I've already discussed social support in this chapter, with particular reference to the lessons in Chapter 6. Invaluable encouragement for self-improvement is available in a work culture with people who trust each other's abilities and intentions and who feel a sense of belonging and interdependency. People are more willing to accept personal responsibility for safety self-management when they believe their coworkers genuinely want them to improve. In fact, people in control of their self-management programs will create a supportive social context for themselves. They will enlist support from supervisors, coworkers, friends, or family members whom they know care. They look around for people who will support their commitment to improve personal safety.

Create a supportive social context for self-management.

Social support can take many forms, and can be either formal or informal. For example, social support can include simple words of encouragement from coworkers, positive feedback from supervisors for personal successes, and/or formal recognition of accomplishments from your workgroup. One way to cultivate opportunities for social support is to display your self-observation data so others can see your accomplishments. You can also create a supportive social environment by signing a behavior-based promise card or devising a behavior-change contract, and then sharing your commitment with coworkers, friends, and family members.

Commitment

As I've detailed in other sources (Geller, 1996a, 1997c), once people make a commitment they encounter internal and external pressures to think and act consistently with their position. That's why people can act themselves into thinking differently, and think themselves into acting differently. This is the Consistency Principle in action, as I discussed earlier in this chapter to explain why it's useful to get people to claim safety as a value. My point here is that you will provide incentive to your self-management project if you make a formal commitment to improve your behavior. You can do this with the following actions:

Behavior-based promise cards and interpersonal contracts are powerful activators.

- signing and publicly displaying a simple promise card indicating your commitment to participate in a self-management process (see Figure 41), or reach a certain self-improvement goal.

- signing a contract with a coworker or friend that specifies the safety-related behavior(s) you wish to improve and includes a brief outline of your plan for accomplishing this goal (see Figure 42).

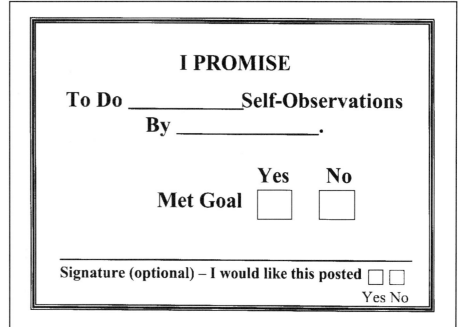

Figure 41. A simple promise card for getting people started on a behavior-based self-management project.

> ### Safe Driving Contract
>
> During the next four weeks I, Joe Peterson, will increase my safety-belt use from 50% to 95%. During that time I will self-observe and record my safety-belt use each day and graph the results. This graph will be placed on my locker. Each day I reach my goal I will reward myself by placing $1 in my vacation fund.
>
> _____   _____
> Signature   Date
>
> During the next four weeks I, Sally Stevens, will help Joe increase his safety-belt use by providing him encouragement on days when he does not reach his goal, and recognize his accomplishments on those days when he does reach his goal. In addition, to show my support of his commitment to increase his safety-belt use, I will set the right example and use my safety belt whenever I travel in a vehicle during this four-week period.
>
> _____   _____
> Signature   Date

Figure 42. Sample safety contract between coworkers for increasing safe driving practices.

You might, for example, draw up a safety contract that specifies the behavior(s) you wish to change, sets a specific behavior-based goal, and indicates a time frame for reaching your goal. The contract should also include a place for both your signature and a person you can call on to help you reach your goal. Then share your contract with someone you'd like to support your behavior improvement efforts. Once you have someone who is willing to support you, you should both sign the contract. A sample contract for promoting safe driving is shown in Figure 42.

A Case Study:

Safety Self-Management of Short-Haul Delivery Drivers

Perhaps the best way to illustrate the potential of safety self-management is to describe some field research my students and I recently completed. The methods and tools used successfully in this study are applicable for a wide range of situations. Please feel free to customize these procedures for your own use.

Our project started in Fall of 1997 when my advanced research students filled out critical behavior checklists while riding in the front passenger seat of a short-haul delivery truck. The drivers were working at a distribution center of a large beverage company.

First define safety-related behaviors to target.

Step 1: Identify and Define Target Behavior

First, eight safe driving behaviors were identified and defined including: a) using running/head lights to increase visibility; b) using a safety belt to reduce injury in case of a traffic crash; c) walking around the truck before departing from a delivery to avoid hitting or running over obstacles; d) maintaining a safe following distance to provide adequate stopping distance in an emergency; e) making a safe lane change by using a turn signal, checking rearview and sideview mirrors, and visually inspecting any blind spots; f) avoiding running yellow and red traffic lights by slowing when approaching traffic lights; g) coming to a complete stop at all stop signs to check for other vehicles; and h) driving at or below the posted speed limit.

The selection of these driving behaviors was based on an analysis of the company's traffic crash reports over the previous two years. Specific definitions of these behaviors were developed and listed on the self-observation checklist depicted in Figure 43.

Vehicle Type: ☐ Truck ☐ Mini Van   Date: _____
(check one)

Instructions. When arriving at each of your accounts or returning to the plant complete the checklist below. Place a tally mark in either "Safe" or "At-Risk" to indicate your driving behaviors when leaving the previous account and traveling to the current account. At the end of the day, total the tally marks for each category ("safe" and "at-risk") and calculate your percent safe using the formula below. Also, total all safe and at-risk behaviors at the bottom of the checklist and calculate an overall percent safe.

SAFE DRIVING BEHAVIOR	Safe	At-Risk	% Safe
Running/Head Lights. Running and/or head lights turned on when vehicle is in motion to increase visibility of vehicle.			
Safety Belt Use. Safety belt used by all vehicle occupants when the vehicle is in motion.			
Walk Around. All sides of vehicle checked for obstacles before leaving previous account.			
Following Distance. Follows the "4 second rule" for trucks and the "2 second rule" for vans.			
Lane Change. Checks mirror, and then signals before changing lanes.			
Traffic Lights. Avoids going through yellow and red lights and/or comes to a complete stop before making a right-on-red.			
Complete Stop at Stop Signs. Comes to a complete stop for 3 seconds before the white line at stop signs.			
Vehicle Speed. Stays at or below the posted speed limit.			
Walk Around. Walks all sides of vehicle to check for obstacles before leaving account.			
TOTAL PERCENT SAFE			

Percent Safe = Number Safe / (Number Safe + Number At-Risk)

COMMENTS:

Thank you for completing this form. Your safe driving efforts are essential for our success in keeping everyone safe on the highway.

Figure 43. Self-Observation Checklist used for a self-management project among 29 short-haul delivery drivers.

The checklist did not include the driver's name, so all self-observation data were anonymous. The participants felt that by keeping their records anonymous, each driver would be more willing to provide an accurate record of driving practices. In fact, some employees were worried about possible retributions for poor safety performance, especially if they happened to have a vehicle crash.

Step 2: Observe Behaviors

During this step of the self-management process a baseline frequency of the eight driving behaviors identified in Step 1 was obtained by my experienced research students. Baseline observations for self-management are normally taken by the individual seeking improvement, but for this research we had an opportunity to obtain baseline measures from independent observers. This provided for more objective data, unbiased by an individual's desire to look good. My students were well-trained and experienced at completing a critical behavior checklist for driving. They rode with the drivers for one day and systematically completed the checklist depicted in Figure 43.

Step 3: Intervene

The drivers decided to target eight behaviors for self-improvement.

The baseline data were shared with the drivers and they decided to self-observe all eight driving behaviors daily over a two-week period. In addition, they decided to post the group data on a graph in the break room on each day of their self-observations. Drivers then self-observed and recorded their driving practices on the self-observation checklist depicted in Figure 43. More specifically, when they arrived at each of the accounts on their sales routes, the drivers recorded whether each behavior on the checklist was performed safely or at-risk while driving to that location.

At the end of each day, the drivers calculated their percent safe score for each driving behavior, and wrote the results in the spaces provided on the self-observation checklist. The total percent safe score was calculated by dividing the total number of safe behaviors by the number of safe plus at-risk behaviors. The self-observation card was then placed in a collection box. Then my research assistants calculated the average percent safe for each behavior (across all 29 drivers), and posted the percentages on a large graph in the break room. This graph of percent safe per behavior was updated daily throughout the two-week self-observation period.

Step 4: Test

To assess the impact of the self-observation intervention, my research assistants went on a second ride-along with the 29 drivers in order to obtain an independent evaluation of their post-intervention driving practices. These ride-alongs occurred about two weeks after the intervention phase ended.

As you can see in Figure 44, the drivers' self-reported percentages of safe driving practices were much higher than the percent safe scores calculated from my students' records taken during the baseline period. But, the good news is that the percent safe scores calculated from by students' post-intervention rides were markedly improved over the percentages they recorded before the drivers' self-observing, recording, and posting intervention. Let's look more closely at these results.

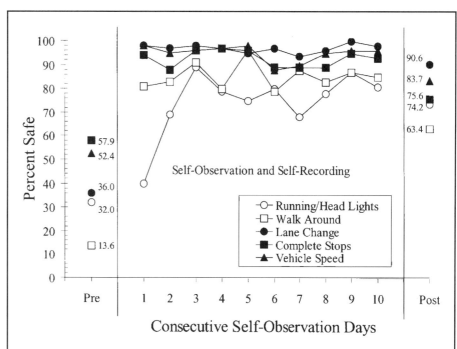

Figure 44. Daily mean percent safe observations of five driving behaviors that improved among 29 short-haul delivery drivers, before intervention (by researchers), during self-management (by drivers), and after the intervention (by researchers).

Five driving behaviors improved markedly after the self-observing, recording and posting interventions.

First, we calculated a total percent safe score for each driver for the pre-intervention (baseline) and post-intervention rides by summing the occurrences of safe and at-risk driving practices across all driving behaviors per session. Then we performed a statistical test to determine if the mean safe driving percentages increased significantly from pre- to post-intervention. The overall increase from 76.7% safe at baseline to 84.6% safe following the intervention was statistically significant at $p<.05$.

Our separate analyses of each driving behavior indicated significant increases in the percentages of these safe behaviors: a) running/headlight use (from 32.0% at baseline to 74.2% following the intervention, b) vehicle walk-arounds (from 16.5% to 50.2%), c) safe lane changes (from 36.0% to 90.6%), d) complete stops at intersections (from 57.9% to 75.6%), and e) trips at or below the posted speed limit (from 52.4% to 83.7%). There was no apparent improvement in safety-belt use (from 75.5% to 72.9%), in maintaining a safe following distance (from 71.8% to 69.0%), nor in avoiding the running of yellow/ red traffic lights (from 86.6% to 83.3%) as a result of the intervention process.

Implications

Overall, our evaluation of this simple intervention process suggests that self-observing, recording, and posting can increase the safe-driving practices of short-haul truck drivers. However, we also found room for further improvement in the truckers' safety and in our intervention process. While there were significant improvements in five of the eight driving behaviors, three of the driving behaviors were unaffected by the self-management process. Also, the drivers' self-reports were substantially safer than indicated in the reports of my research assistants. It's likely some drivers exaggerated their safety when completing their own checklists. It's also likely the self-management process was not long enough for drivers to develop safe driving habits. Therefore some safe behaviors reverted to at-risk behaviors after the intervention period ended.

Our self-management intervention was too brief, and targeted too many behaviors.

Besides being too brief, the intervention process probably targeted too many behaviors at once. The drivers may have had difficulty remembering so many driving behaviors, or they may have focused on only a few of the driving behaviors and neglected others. When recording their performance it's likely many employees had to guess whether they were safe or at risk on some driving behaviors. It would have been more effective if each driver had reviewed his baseline data to define two or three driving behaviors to target during the self-management intervention phase. This is, in fact, the approach discussed earlier in this chapter as the way to DO IT. Use initial observing and recording to pinpoint critical behaviors to improve, and then target only these few behaviors during intervention.

We could have used several additional techniques.

Another weakness of this field study was the limited use of intervention techniques. It is gratifying that only self-observing, recording, and posting of certain driving behaviors lead to rapid and significant improvement. However, a number of additional self-management techniques could have been used to obtain greater beneficial impact. The techniques discussed earlier, including activator management, self-statements, imagery, goal-setting, commitment, self-rewards, and social support could have been readily adapted to this driving situation. Unfortunately, we didn't have the opportunity to teach self-management principles and procedures to these drivers and help them customize their own self-improvement process. Such instruction and intervention design would require substantial time off-the-job, and these drivers' supervisors are not yet ready to make such a commitment to safety self-management. Perhaps these initial results will help turn that around.

Use the DO IT process to design your own self-management project.

Figure 45 provides an opportunity for you or the participants in an education/training program to design a safety self-management project. The exercise takes the participant through the four stages of DO IT — Define, Observe, Intervene, and Test. After answering the procedural questions, the participant has an action plan for initiating a safety self-management project. You could ask the participants to actually conduct their plans, and to give progress reports at follow-up group/ team meetings.

If actual implementation is your aim, then it's important to review each participant's proposal for practicality and adherence to the principles and procedures discussed in this chapter. In this regard, it could be very useful for the participants to explain their self-management DO IT plan to their education/training group. This provides an opportunity for team members to provide constructive feedback, as well as to offer social support. And, of course, when an individual presents a personal improvement plan in front of his or her coworkers, a strong sense of commitment develops, increasing the likelihood the self-management process will be implemented.

Directions: Please show your operational understanding of the important **DO IT** process by developing a self-management action plan that uses this process. Your action plan can be specified by defining procedures for accomplishing each component of the **DO IT** process: **D** = Define the target behavior, **O** = Observe occurrences of the target behavior, **I** = Intervene to change the frequency of the target behavior in desired directions, and **T** = Test the impact of the intervention strategy by continuing to record occurrences of the target behavior.

1. **Define** [What personal safety-related behavior do you want to decrease or increase? Be sure to describe the behavior in observable, measurable terms.]

   _____
   _____
   _____

2. **Observe** [How will you observe the target behavior prior to intervention to develop a baseline measure of the behavior? Where, how, and how often? What kind of checklist will you use to record occurrences of the target behavior?]

   _____
   _____
   _____
   _____

3. **Intervene** [What self-management technique(s) will you use to influence the target behavior?]

   _____
   _____
   _____

4. **Test** [How will you collect data or information to compare to that collected during the baseline measurement period? How will you make comparisons to test the impact of your intervention(s)? How will you continually monitor the progress of your intervention to ensure you maintain the desired level of behavior change?]

   _____
   _____
   _____
   _____

Figure 45. Developing a DO IT Process for Safety Self-Management.

Chapter 7 — Teach and Support Safety Self-Management

In Conclusion

This chapter can be considered a summary of this entire book. It shows how the lessons in each chapter relate to building a Total Safety Culture. Safety self-management means taking personal responsibility for doing the right things to prevent injury to both self and others. It means going beyond the duties defined by an accountability system. When you're out there by yourself, with no one to hold you accountable, do you choose the safe behavior or a more efficient or convenient at-risk behavior? In other words, do you take a calculated risk when you think you can get away with it? Safety self-management techniques can help you choose the safe way consistently, but you need to choose to use these techniques. And this takes commitment and personal responsibility.

Safety self-management requires a paradigm shift from accountability to personal responsibility.

We need to help people transition from an accountability to a responsibility mindset. This is not easy in some work cultures, especially when it comes to safety. Traditionally, safety promotion has been top-down. Safety-related conversations often resemble an adult-child confrontation in which one person holds another person accountable for at-risk behavior. Such interaction is interpreted as a "gotcha" process, backed by a government or corporate rule or regulation. After all, this is the way governments manage the behavior of the masses — they pass a law and enforce it.

Reflect how you feel when your behavior is motivated by a law or regulation. You don't necessarily believe the law is important for you, but you obey it anyway in order to avoid a penalty. How do you feel in this situation? Well, you might feel controlled and not empowered. Following a rule doesn't make you feel important, nor does it improve your self-esteem. And it doesn't increase your sense of belonging. It might even decrease one or more of these person states if the circumstances remind you of the top-down control. This is opposite of what is needed to build personal responsibility for safety throughout a work culture.

The DO IT process is self-rewarding once it gets started.

Safety self-management involves people in behavior-based safety for their own self protection. It requires a little extra effort in **D**efining target behaviors to improve, **O**bserving and recording the current frequency of the behaviors, **I**ntervening to increase safe and decrease at-risk behaviors, and then **T**esting the impact of your intervention. This DO IT process is straightforward and self-rewarding once it gets going. But this basic self-improvement process is difficult to implement in some work contexts. Most of the practical lessons in this book focus on ways to develop a work context that builds personal responsibility for safety and facilitates the application of techniques for safety self-management.

Seven Self-management intervention strategies were reviewed in this chapter. Researchers have shown each of these techniques effective at improving target behaviors beyond levels obtained during initial self-observing and recording. These intervention techniques include the following: a) managing environmental activators or conditions antecedent to a target behavior; b) developing and using self-statements to encourage desired behavior and discourage undesirable behavior, including self-instructions, beliefs, and interpretations; c) picturing situations in your mind's eye to direct and motivate target behavior; d) administering self-rewards strategically to support desired behaviors, including the completion of certain self-management processes; e) setting Specific, Motivational, Achievable, Recordable, and Trackable goals regarding both the completion of specified self-management techniques and the achievement of a certain level of behavioral improvement; f) making an explicit voluntary commitment to complete particular aspects of a self-improvement project and/or to reach a designated level of behavioral improvement; and g) enlisting the support of one of more friends or coworkers for social encouragement and support.

These seven intervention approaches to self-management are listed in Figure 46 as a convenient reference. I bet you can design an improvement strategy for each of these approaches. The most effective self-management projects use intervention techniques for each of the categories in Figure 46, but don't try to do too much at once. Start small and build on your successes.

As in behavior-based shaping, each successive improvement develops more skill, self-efficacy, personal control, and optimism, preparing you for a higher level of accomplishment. Begin with the vision of a Total Safety Culture, but focus your daily activities on goals that approximate the ultimate outcome. And remember to celebrate the achievement of process goals. Recognize and self-reward your small wins. Safety self-improvement is a journey, not a destination.

❑ *Activator Management* — placing activators such as a reminder message to direct safe behavior.

❑ *Self-Statements* — using self instructions, beliefs, and interpretations to activate safe behavior.

❑ *Mental Imagery* — forming a mental picture of safe behavior as well as directive activators and motivational consequences.

❑ *Self-Reward* — giving oneself a self-determined reward after reaching a self-determined safety goal.

❑ *Goal-Setting* — specifying recordable and trackable safety improvements to achieve in a given period.

❑ *Commitment* — writing down a personal promise to work toward achieving a SMART safety goal.

❑ *Social Support* — enlisting support from work supervisors, coworkers, friends, and family to encourage the safe behavior targeted.

Figure 46. Seven intervention strategies facilitate safety self-management.

Beyond Safety Accountability

Bibliography

American Heritage Dictionary (1991). Second College Edition. New York: Houghton Mifflin Company.

Bird, F. E., Jr., & Germain, G. L. (1997). *The Property Damage Accident: The Neglected Part of Safety.* Loganville, GA: Institute Publishing, Inc.

Brown, Jr., H. J. (1991). *Life's Little Instruction Book.* Nashville, TN: Rutledge Hill Press.

Brown, Jr., H. J. (1993). *Life's Little Instruction Book, Volume II.* Nashville, TN: Rutledge Hill Press.

Carnegie, D. (1936). *How to Win Friends and Influence People.* New York: Simon & Schuster, Inc.

Cialdini, R. B. (1993). *Influence: Science and Practice* (Third Edition). New York: Harper Collins College Publishers.

Cook, J., & Wall, T. (1980). New Work Attitude Measures of Trust, Organizational Commitment and Personal Need Non-Fulfillment. *Journal of Occupational Psychology, 53,* 39-52.

Deming, W. E. (1985). Transformation of Western Style of Management. *Interfaces,* 13(3), 6-11.

Freedman, J. L. (1965). Long-Term Behavioral Effects of Cognitive Dissonance. *Journal of Experimental Social Psychology,* 1, 145-155.

Gardner, H. (1993). *Multiple Intelligences.* New York: Basic Book.

Geller, E. S. (1996a). *The Psychology of Safety: How to Improve Behaviors and Attitudes on the Job.* Boca Raton, FL: CRC Press.

Geller, E. S. (1996b) The Truth About Safety Incentives. *Professional Safety,* 41(10), 34-39.

Geller, E. S. (1997a). *Actively Caring for Safety: The Psychology of Injury Prevention.* (12-tape cassette series). Blacksburg, VA: Safety Performance Solutions.

Geller, E. S. (1997b). Key Processes for Continuous Safety Improvement: Behavior-Based Recognition and Celebration. *Professional Safety,* 42(10), 40-44.

Geller, E. S. (1997c). *Understanding Behavior-Based Safety: Step-by-Step Methods to Improve Your Workplace.* Neenah, WI: J. J. Keller & Associates.

Geller, E. S. (1998). *Building Successful Safety Teams: Together Everyone Achieves More.* Neenah, WI: J. J. Keller & Associates.

Geller, E. S., Lehman, G. R., & Kalsher, M. J. (1989). *Behavior Analysis Training for Occupational Safety.* Newport, VA: Make-A-Difference, Inc.

Geller, E. S., Winett, R. A., & Everett, P. B. (1982). *Preserving the Environment: New Strategies for Behavior Change.* New York: Pergamon Press.

Goldstein, A. P., & Krasner, L. (1987). *Modern Applied Psychology.* New York: Pergamon Press.

Goleman, D. (1993). *Emotional Intelligence.* New York: Bantam Books.

Greene, B. F., Winett, R. A., Van Houten, R., Geller, E. S., & Iwata, B. A. (1987) (Eds.). *Behavior Analysis in the Community: Readings from the Journal of Applied Behavior Analysis.* University of Kansas, Lawrence, KS.

Heinrich, H. W. (1931). *Industrial Accident Prevention.* New York: McGraw-Hill.

James, W. (1890). *Principles of Psychology.* New York: Doves.

Latané, B., & Darley, J. M. (1970). *The Unresponsible Bystander: Why Doesn't He Help?* New York: Appleton-Century-Crofts.

Norman, D. A. (1988). *The Psychology of Everyday Things.* New York: Basic Books.

Peck, M. Scott. (1987). *The Different Drum: Community Making and Peace.* New York: Simon & Schuster, Inc.

Rees, F. (1997). *Teamwork from Start to Finish.* San Francisco, CA: Jossey-Bass, Inc.

Seligman, M. E. P. (1975). *Helplessness.* San Francisco, CA: W. H. Freeman.

Skinner, B. F. (1938). *The Behavior of Organisms: An Experimental Analysis.* Acton, MA: Copley Publishing Group.

Skinner, B. F. (1950). Are Theories of Learning Necessary? *Psychological Review,* 57, 193-216.

Skinner, B. F. (1953). *Science and Human Behavior.* New York: Macmillan.

Skinner, B. F. (1971). *Beyond Freedom and Dignity.* New York: Alfred A. Knopf.

Skinner, B. F. (1974). *About Behaviorism.* New York: Alfred A. Knopf.

Steiner, I. D. (1970). Perceived Freedom. In L. Berkowitz (Ed.), *Advances in Experimental Social Psychology,* (Vol. 5). New York: Academic Press.

Watson, D. L., & Tharp, R. G. (1993). *Self-Directed Behavior: Self-Modification for Personal Adjustment* (Sixth Edition). Pacific Grove, CA: Brooks/Cole Publishing Company.

White, R. W. (1959). Motivation Reconsidered: The Concept of Competence. *Psychological Review*, 66, 297-333.

Beyond Safety Accountability

Glossary: The Language of Safety Accountability and Responsibility

Teaching and applying a new technology requires learning a new language. This module introduces the language of behavior-based safety as it relates to safety accountability and personal responsibility. Sometimes the meaning of words used to describe the principles and methods of behavior-based accountability and responsibility are different than the meanings implied in everyday use of the same words. Often a more precise definition of certain words is needed in order to communicate relevant principles or methods.

The following words are key to understanding, teaching, and using the practical methods presented in this book. The brief and specific definitions reflect the behavior-based perspective. When applying behavior-based principles and methods to safety accountability and personal responsibility, it's a good idea to use these words consistently with the definitions given here. This puts everyone on the same wavelength and improves communication.

Accountability: a system, process, or contingency that defines desired outcomes from individual or group performance.

Activator: a stimulus or antecedent event that signals the occurrence or the absence of a particular behavior or set of behaviors. Activators direct behavior.

Active Listening: listening attentively to understand another person's or group's perspective.

Actively Caring: going beyond one's normal routine to look out for the safety or health of another person.

Applied Behavior Analysis: the application of techniques developed from experimental behavior analysis to evaluate behavior in real-world settings, to increase the occurrence of desirable behavior, and to decrease the occurrence of undesirable behavior.

At-Risk Behavior: an activity or behavior that puts a person in some degree of jeopardy to be injured.

Attitude: a learned belief, disposition, or mindset regarding a particular person, event, situation, or set of circumstances.

Attribution: assigning a cause to an event, to one's own behavior, or to another person's behavior.

Baseline: the condition or time period in which no intervention is implemented.

Glossary

Behavior Sampling: observing another person doing a particular activity in order to analyze the situation and provide feedback for continuous improvement.

Behavior: any observable, measurable activity of an organism.

Behavioral Deficit: a desirable target behavior an individual seeks to increase in frequency, duration, or intensity.

Behavioral Excess: an undesirable target behavior an individual seeks to decrease in frequency, duration, or intensity.

Behavioral Outcome: the result or condition following the occurrence of one or more behaviors.

Behavior-Based Incentive: an announcement or promotion that a certain reward will be distributed after a designated number of behaviors are performed or process activities completed.

Behavior-Based: a focus on the overt, observable actions of people, as well as the objective environmental or cultural factors that influence behavior.

Behavior-Reward Contingency: a designated relationship between one or more behaviors and the rewarding consequence, which defines the delivery system for a behavior-based incentive program.

Belonging: the extent to which a person feels part of a team or large group, or feels a sense of interdependency or interconnectedness with others.

Calculated Risk: an at-risk behavior that is intentional, usually in an attempt to be comfortable, convenient, or efficient.

Candor: straightforwardness and frankness of expression; freedom from prejudice.

Caring: showing concern or interest about what happens.

Character: the combined moral or ethical structure of a person or group; integrity; fortitude.

Coaching: a process whereby an individual observes and analyzes another person's behavior and then communicates findings in order to improve the performance of that person.

Cognition: internal or covert verbal behavior including thinking, imagining specific situations or behaviors, talking to oneself, or recalling events of the past.

Commitment: a written or oral promise to perform a particular behavior or achieve a certain process or outcome goal.

Common Sense: ideas, principles, concepts, or procedures based on personal experience, which is necessarily selective and biased.

Communication: exchange of information or opinion by speech, writing, or signals.

Conformity: a change in behavior or attitude activated by a desire to follow the actions, beliefs, or values of other people.

Conscious Competence: aware of or thinking about one's correct (or safe) behavior.

Conscious Incompetence: aware of or thinking about one's incorrect (or at-risk) behavior, as when taking a calculated risk.

Consequence: the stimulus or event occurring immediately after a behavior.

Consistency: agreement among successive acts, ideas, or events.

Context: the surroundings and circumstances in which a particular event or behavior occurs.

Contingency: a relationship between a behavior and a consequence such that the consequence is presented if and only if the behavior occurs.

Corrective Action: a set of behaviors or intervention plan to change an environmental, behavioral, or person-state factor that could contribute to an injury if not corrected.

Countercontrol: a side-effect of punishment or a top-down mandate whereby people feel controlled and react negatively, performing undesirable behavior to reassert their freedom or individuality.

Credit Economy System: a behavior-based incentive program whereby people earn credits or points for various target behaviors, and later exchange them for designated prizes.

Critical Behavior: a behavior that needs to occur or needs to be avoided in order to prevent property damage or personal injury.

Critical Behavior Checklist: a list of specific behavioral definitions used to implement behavior-based observation and feedback.

Culture: the combined learned behaviors, beliefs, attitudes, and values characteristic of individuals in a group, organization, community, or society as a whole.

Dependent Perspective: to view oneself as dependent on others for achieving certain goals or objectives.

DO IT Process: a four-step behavior-based process for continuous safety improvement, with D = define target behavior(s), O = observe occurrences of the target behavior(s), I = intervene to improve the target behavior(s), and T = test the impact of the behavior-based intervention on the target behavior(s).

Domain of Accountability: the environmental territory or sphere of activity which a person must answer to with regard to a particular evaluation process.

Domain of Responsibility: the environmental territory or sphere of activity which a person feels empowered to influence.

Education: a process whereby people learn the principles or rationale underlying a particular method, program, or process.

Emotional Intelligence: the ability to take a broad and long-term perspective; willingness to delay an immediate and certain reward in order to receive a more desirable but less certain one later.

GLOSSARY

Empathic Listening: listening that attempts to identify with and understand another person's feelings, intentions, or motives.

Empowerment: the extent to which a person believes he or she can accomplish a particular assignment or challenge; it is influenced by levels of self-efficacy, personal control, and optimism.

Environment: the physical context in which we behave; in the workplace it includes such items as buildings, machines, tools, materials, personal protective equipment, and temperature.

Environmental Checklist: a list of factors in the work environment, including equipment, tools, and furniture which could influence the occurrence or prevention of an injury.

Ergonomics: the study of relationships between human anatomy or physiology, the demands of a particular task or piece of equipment, and the environment in which the behavior occurs.

Error: at-risk or unsafe behavior that is unintentional.

Expectancy: a person state that leads an individual to anticipate the occurrence of a particular environmental event.

Extrinsic Consequence: a stimulus or event (reward or penalty) given after the occurrence of a behavior in order to motivate its increase or decrease.

Extrinsic Motivation: when a person participates on a project or performs certain behaviors in order to receive an external consequence or reward.

Feedback: a natural or contrived event that motivates behavior when it follows it (as a consequence), or directs behavior when it precedes it (as an activator).

Frequency: a dimension of behavior — specifically the number of times a behavior occurs in a certain time period.

Fundamental Attribution Error: the tendency to attribute other people's behavior to inside person factors and to minimize the importance of outside situational factors.

Goal-Setting: a technique to motivate behavior whereby specific, measurable, and achievable results of a particular activity or process are defined; then progress toward the goal is recorded and tracked so participants can see their accomplishments and predict when the goal will be reached.

Group: two or more people who interact with one another, perceive themselves as a group, and are interdependent.

Habituation: a decrease in responding to repeated presentations of the same stimulus or activator.

Heinrich's Ratio: an estimate proposed in 1931 by H. W. Heinrich that for every major injury, 29 minor injuries and 300 near-miss incidents occur.

Imagery: a cognitive or thinking process in which a mental picture of a sensory or perceptual experience is created.

Incentive: an announcement of the availability of a reward for the performance of certain behavior.

Incident Analysis Team: a work group that conducts fact-finding evaluations of "near-miss" reports and injuries.

Independent Perspective: to view oneself as personally responsible for achieving certain goals or objectives.

Interdependence: the degree of influence two people have over each other, including the quantity and quality of activities in which they jointly engage.

Interdependent Perspective: to view oneself as a member of a larger community of people, working together in win/win relationships to achieve synergy or social harmony.

Internal Control: acquiring self-discipline or the ability to perform in a certain way because of an internal mental script rather than for external activators and consequences.

Interpersonal Trust: trust between individuals developed through one-on-one communication, caring, candor, consistency, commitment, consensus, and character.

Intervention: anything an individual or group does in an attempt to improve behaviors, attitudes, or environmental conditions.

Intrapersonal Intelligence: the ability to motivate one's self and build one's self-esteem, self-efficacy, personal control, optimism, or sense of belonging.

Job-Specific Checklist: a critical behavior checklist (CBC) that targets the safe and at-risk work practices of a particular job.

Learned Helplessness: a person state in which people believe they cannot deal with an ongoing challenge, frustration, or distress, and they give up trying; it can lead to depression.

Learning: any relatively permanent change in behavior (or behavioral potential) as the result of experience.

Mental Script: words we say to ourselves which direct or activate our behavior (as an intention), or which provide a consequence to motivate us to increase or decrease the occurrence of a behavior (as in self-reinforcement or self-punishment, respectively).

Mission Statement: writing that articulates the overall purpose or intent of an organization or a team.

Mistake: at-risk behavior that is unintentional and results from poor judgment due to a deficit in knowledge, skill, or experience.

Motivation: when people perform to obtain expected internal or external consequences.

Natural Consequence: a pleasant or unpleasant stimulus or event inherent or intrinsic to a particular behavior or activity which motivates the person to continue or stop performing the behavior.

Near Miss: an incident that did not result in an injury but would have under slightly different circumstances.

GLOSSARY

Objective Statement: an opinion or inference based entirely on what a person sees in the physical world.

Observation and Feedback Team: a work group that develops, implements, evaluates, and refines behavior-based observation and feedback procedures.

Optimism: the extent to which a person expects to obtain desired outcomes from a certain activity or set of behaviors.

Organizational Culture: the dominant pattern of basic assumptions, perceptions, thoughts, feelings, and attitudes held by members of an organization.

Outcome Goal: an anticipated result of individual or group effort, specifically defined as a level of achievement to work for through process activities.

Outcome-Based Incentive: an announcement or promotion that a particular reward will be distributed after reaching a particular outcome, like having no injuries over a six-month period.

Ownership: the extent that individuals, groups, or organizations perceive a program, project, or process belongs to them and is under their direct control. Greater ownership leads to greater involvement because "owners" have a personal stake in the results of a program, project, or process.

Percent Safe Behavior: the number of safe behaviors observed divided by the total number of observations (safe plus at-risk behaviors), and then multiplied by 100 to give a percentage.

Percent Safe Employees: the total number of employees observed who, according to the critical behaviors on a checklist, performed everything safely, divided by the total number of employees observed, and then multiplied by 100 to give a percentage.

Perception: the process through which we select, organize, and interpret input from our sensory receptors; the subjective interpretation of what we see, hear, smell, feel, and taste.

Perception Survey: a questionnaire used to assess people's feelings, opinions, perceptions, or attitudes about specific events, situations, or circumstances.

Person State: a temporary and changeable sensation or internal feeling that influences a person's behavior.

Person Trait: a permanent and stable personality characteristic of an individual that influences his or her behavior.

Personal Control: the extent to which a person believes he or she has direct individual control of the factors needed to achieve a particular outcome.

Person-Based: a focus on subjective and internal feeling states that can influence attitude and behavior.

Person Factors: internal aspects of people influencing their behavior, including knowledge, skills, abilities, attitudes, feelings, mood states, and values.

Prejudice: the positive or negative evaluations or judgments of members of a group based primarily on membership in the group and not on the particular characteristics of individual members.

Principle: a basic truth, law, or assumption established from systematic scientific research.

Process: procedures or methods that are followed and continuously improved in order to achieve certain goals.

Process Goal: a certain amount of effort anticipated by an individual or group to accomplish a desired outcome, and defined specifically with regard to number of behaviors performed or individuals participating.

Psychology: the scientific study of behavior-based (external) and person-based (internal) aspects of people.

Punisher: a stimulus or event which results in a decrease in the occurrence of a behavior when the stimulus or event is presented immediately following that behavior. [Throughout this book I use less technical terms like "penalty" or "negative consequence." Unlike these more popular terms, however, a consequence is not a punisher if it does not decrease the frequency of the response it follows.]

Punishment: a procedure used to decrease the frequency of a behavior by introducing an unpleasant stimulus immediately following the response.

Purpose: an ultimate outcome or vision that can motivate the development of specific programs or processes and the performance of various behaviors.

Recognition: praise or interpersonal approval given to a person for desirable behavior.

Research: application of the scientific method rather than common sense or intuition to acquire knowledge and decide on intervention approaches.

Response Generalization: the occurrence of different but related behaviors in the same situation. For example, if one safety behavior is rewarded in a particular work setting, other similar safety behaviors might increase in frequency.

Responsibility: the extent to which a person feels empowered (i.e., capable, in control, and optimistic) to achieve a particular goal, outcome, or vision.

Reward: a stimulus or event presented immediately following a certain behavior in an attempt to increase the occurrence of that behavior.

Root Cause: the underlying critical factor presumed to influence a particular incident, like property damage or personal injury.

Rule-Governed Behavior: Behavior controlled by a verbal statement (a rule) about a contingency between the behavior and a consequence.

Safe Behavior: an activity or behavior that reduces or eliminates the possibility of personal injury.

Safety Share: an activator to increase safety awareness and promote an achievement perspective, whereby individuals at a group meeting are asked to state something they've done for safety over a prior time period or to state something they intend to do for safety within a future period.

Glossary

Safety Triad: the three domains needing attention in order to cultivate a Total Safety Culture — environment, behavior, and person.

Scientific Method: an approach to knowledge acquisition that emphasizes empirical rather than intuitive processes, systematic and controlled observation of operationally defined phenomena, data collection using reliable and valid measurement procedures, and objective reporting of results.

Self-Efficacy: the extent to which a person believes he or she can successfully perform a particular task or behavior.

Self-Esteem: one's self concept; the extent to which a person feels good about himself or herself.

Self-Instructions: self-statements given to make it more likely a target behavior will occur in a specific situation.

Self-Management: application of behavior analysis techniques by an individual to change one or more personal behaviors.

Self-Monitoring: a type of direct observation and data collection process whereby an individual observes and records his or her own behavior as they occur.

Self-Punishment: the delivery of an external or internal penalty to oneself after making an error or failing to reach a goal, which can decrease the occurrence of the behavior it follows and depreciate self-esteem, self-efficacy, and feelings of personal control.

Self-Reward: a component of self-management whereby an individual gives him or herself an external or internal positive consequence for performing a particular target behavior or achieving a designated goal.

Shaping: the procedure of reinforcing successive approximations of a behavior until a terminal or goal behavior is reached.

Slip: at-risk behavior that is unintentional, and usually results from a cognitive failure or "brain cramp."

Social Support: the comfort, recognition, approval, or encouragement available from other people, including friends, family, and coworkers.

Stimulus Generalization: similar behaviors are performed in situations different from the original training environment. For example, if a certain safety behavior is rewarded in one setting, that same behavior may occur in different settings even though rewards are not available in these settings.

Stimulus: An environmental event that can be detected by one of the senses.

Successive Approximation: a behavior in the shaping process that looks more like the target behavior. The shaping process starts with rewarding the first approximation to the target behavior. After this behavior is strengthened through positive reinforcement, only a closer approximation is rewarded. This process continues until the target behavior is performed.

Synergy: the action of group or team members to achieve results which each is individually incapable of achieving.

System: a group of interacting, interrelated, and interdependent elements forming a complex whole.

Systems Thinking: considering situations and circumstances from a broad-based, long-term, and interdependent perspective.

Target Behavior: any observable, measurable activity of an individual that is the focus of behavior-based analysis or intervention.

Team: a collection of two or more individuals with complimentary skills who are committed to a common mission, performance goals, methods, and procedures for which they hold each other mutually accountable.

Three-Term Contingency: the activator directing behavior (A), the behavior (B), and the motivating consequence (C).

Total Recordable Injury Rate (TRIR): the total number of injuries reported by employees that reach a designated level of severity, like those requiring treatment by a physician.

Total Safety Culture: the ultimate vision of a safety-improvement mission — a work culture in which everyone feels responsible for safety and pursues it on a daily basis; safety is not a priority that gets shifted according to situational demands but is a value linked to all situational priorities.

Training: a process whereby people learn the specific behaviors or step-by-step procedures required to complete a certain task, carry out a particular program, or achieve a certain goal or objective.

Trust: confidence in the integrity, ability, character, and truth of a person or thing.

Unconscious Competence: performing a correct (or safe) behavior automatically or out of habit without thinking about it.

Unconscious Incompetence: performing an incorrect (or at-risk) behavior automatically or out of habit, without thinking about it.

Value: a deep-seated internal principle, standard, or belief that influences attitude and behavior; it provides the foundation for the enduring uncompromising rules a person attempts to follow consistently.

Vision: an ultimate outcome or consequence such as "an injury-free workplace" which motivates goal-setting and the development and implementation of processes, programs, or behaviors.

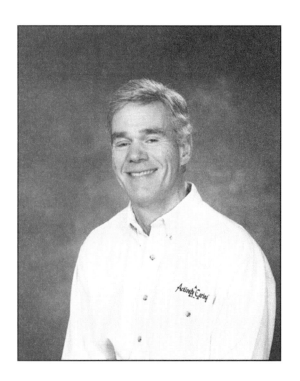

E. Scott Geller, Ph.D., and professor of psychology at Virginia Polytechnic Institute and State University for almost 30 years, has published more than 200 research articles in professional journals and more than 75 articles in *Industrial Safety and Hygiene News*. Recently he authored two books on the psychology of safety; one for safety professionals (*The Psychology of Safety: How to improve behaviors and attitudes on the job*), and the other for the general public (*Working Safe: How to help people actively care for health and safety*).

A senior partner with Safety Performance Solutions, Dr. Geller has pioneered the application of behavioral science toward solving safety, health, and environmental problems in corporate and community settings. He has worked with numerous public and private agencies and organizations to tailor training programs and intervention processes based on behavior- and person-based principles and methods.

Dr. Geller has been the principal investigator for more than 70 research grants and contracts from a variety of corporations and government agencies including the National Science Foundation; The National Institute for Occupational Safety and Health; The U.S. Department of Health, Education and Welfare; the U.S. Department of Energy; The U.S. Department of Transportation; the National Highway Traffic Safety Administration; the National Institute on Alcohol Abuse and Alcoholism; and the Centers for Disease Control.

Dr. Geller is a Fellow of the American Psychological Association, the American Psychological Society, and the World Academy of Productivity and Quality Sciences. He is past Editor of the *Journal of Applied Behavior Analysis* (1989-1992), current Associate Editor of *Environment and Behavior* (since 1983), and current Consulting Editor for *Behavior and Social Issues*, the *Behavior Analyst Digest*, and the *Journal of Organizational Behavior Management*.

Professor Geller was the 1982 recipient of the American Psychological Association, Division Two Teaching Award in the four-year College/University Division. In 1983 he received the Virginia Tech Alumni Teaching Award, and was elected to the Virginia Tech Academy of Teaching Excellence; in 1990 he was honored with the all-university Sporn Award for distinguished teaching of freshmen-level courses. In 1998, the Society for the Advancement of Behavior Analysis honored Dr. Geller with the Award for Effective Presentation of Behavior Analysis in the Mass Media.

Government Institutes Mini-Catalog

PC #	ENVIRONMENTAL TITLES	Pub Date	Price*
629	ABCs of Environmental Regulation	1998	$65
672	Book of Lists for Regulated Hazardous Substances, 9th Edition	1999	$95
4100	CFR Chemical Lists on CD ROM, 1999-2000 Edition	1999	$125
512	Clean Water Handbook, Second Edition	1996	$115
581	EH&S Auditing Made Easy	1997	$95
673	E H & S CFR Training Requirements, Fourth Edition	2000	$99
825	Environmental, Health and Safety Audits, 8th Edition	2001	$115
548	Environmental Engineering and Science	1997	$95
643	Environmental Guide to the Internet, Fourth Edition	1998	$75
820	Environmental Law Handbook, Sixteenth Edition	2001	$99
688	EH&S Dictionary: Official Regulatory Terms, Seventh Edition	2000	$95
821	Environmental Statutes, 2001 Edition	2001	$115
4099	Environmental Statutes on CD ROM for Windows-Single User, 1999 Ed.	1999	$169
707	Federal Facility Environmental Compliance and Enforcement Guide	2000	$115
708	Federal Facility Environmental Management Systems	2000	$99
689	Fundamentals of Site Remediation	2000	$85
515	Industrial Environmental Management: A Practical Approach	1996	$95
510	ISO 14000: Understanding Environmental Standards	1996	$85
551	ISO 14001: An Executive Report	1996	$75
588	International Environmental Auditing	1998	$179
518	Lead Regulation Handbook	1996	$95
608	NEPA Effectiveness: Mastering the Process	1998	$95
582	Recycling & Waste Mgmt Guide to the Internet	1997	$65
615	Risk Management Planning Handbook	1998	$105
603	Superfund Manual, 6th Edition	1997	$129
685	State Environmental Agencies on the Internet	1999	$75
566	TSCA Handbook, Third Edition	1997	$115
534	Wetland Mitigation: Mitigation Banking and Other Strategies	1997	$95

PC #	SAFETY and HEALTH TITLES	Pub Date	Price*
697	Applied Statistics in Occupational Safety and Health	2000	$105
547	Construction Safety Handbook	1996	$95
553	Cumulative Trauma Disorders	1997	$75
663	Forklift Safety, Second Edition	1999	$85
709	Fundamentals of Occupational Safety & Health, Second Edition	2001	$69
612	HAZWOPER Incident Command	1998	$75
662	Machine Guarding Handbook	1999	$75
535	Making Sense of OSHA Compliance	1997	$75
718	OSHA's New Ergonomic Standard	2001	$95
558	PPE Made Easy	1998	$95
683	Product Safety Handbook	2001	$95
598	Project Mgmt for E H & S Professionals	1997	$85
658	Root Cause Analysis	1999	$105
552	Safety & Health in Agriculture, Forestry and Fisheries	1997	$155
669	Safety & Health on the Internet, Third Edition	1999	$75
668	Safety Made Easy, Second Edition	1999	$75
590	Your Company Safety and Health Manual	1997	$95

Government Institutes
4 Research Place, Suite 200 • Rockville, MD 20850-3226
Tel. (301) 921-2323 • FAX (301) 921-0264
Email: giinfo@govinst.com • Internet: http://www.govinst.com

Please call our customer service department at (301) 921-2323 for a free publications catalog.

CFRs now available online. Call (301) 921-2355 for info.

*All prices are subject to change. Please call for current prices and availablity.

Government Institutes Order Form

4 Research Place, Suite 200 • Rockville, MD 20850-3226
Tel (301) 921-2323 • Fax (301) 921-0264
Internet: http://www.govinst.com • E-mail: giinfo@govinst.com

4 EASY WAYS TO ORDER

1. **Tel:** (301) 921-2323
   Have your credit card ready when you call.

2. **Fax:** (301) 921-0264
   Fax this completed order form with your company purchase order or credit card information.

3. **Mail:** **Government Institutes Division**
   ABS Group Inc.
   P.O. Box 846304
   Dallas, TX 75284-6304   USA
   Mail this completed order form with a check, company purchase order, or credit card information.

4. **Online:** Visit http://www.govinst.com

PAYMENT OPTIONS

❑ **Check** *(payable in US dollars to **ABS Group Inc. Government Institutes Division**)*

❑ **Purchase Order** *(This order form must be attached to your company P.O. Note: All International orders must be prepaid.)*

❑ **Credit Card** ❑ VISA ❑ MasterCard ❑ AMERICAN EXPRESS

Exp. ___ /___

Credit Card No. _____

Signature _____

(Government Institutes' Federal I.D.# is 13-2695912)

CUSTOMER INFORMATION

Ship To: (Please attach your purchase order)

Name _____
GI Account # *(7 digits on mailing label)* _____
Company/Institution _____
Address _____
(Please supply street address for UPS shipping)
_____

City _____ State/Province _____
Zip/Postal Code _____ Country _____
Tel ( ) _____
Fax ( ) _____
E-mail Address _____

Bill To: (if different from ship-to address)

Name _____
Title/Position _____
Company/Institution _____
Address _____
(Please supply street address for UPS shipping)
_____

City _____ State/Province _____
Zip/Postal Code _____ Country _____
Tel ( ) _____
Fax ( ) _____
E-mail Address _____

Qty.	Product Code	Title	Price

30 DAY MONEY-BACK GUARANTEE
If you're not completely satisfied with any product, return it undamaged within 30 days for a full and immediate refund on the price of the product.

Subtotal _____
MD Residents add 5% Sales Tax _____
Shipping and Handling (see box below) _____
Total Payment Enclosed _____

SOURCE CODE: BP03

Shipping and Handling
Within U.S.:
1-4 products: $6/product
5 or more: $4/product
Outside U.S.:
Add $15 for each item (Global)

Sales Tax
Maryland 5%
Texas 8.25%
Virginia 4.5%